Nutrition Security for Planetary Health

Using the lens of nutrition security and equity for living beings and living systems of our planet, *Nutrition Security for Planetary Health* takes an integrated approach that not only delineates the antecedents of the multifaceted environmental crises—but offers solutions that include the extensive co-benefits of whole plant foods nutrition as the foundational dietary pattern for improving planetary health. Overlooked yet inseparable problems connecting food systems to the transgression of our planetary boundaries, chronic disease, and zoonotic disease are discerned. Bolstering nature and biodiversity is emphasized throughout including regenerative agroecology and native plant foods and ecosystems, utilizing traditional and Indigenous wisdom, rewilding—and community science. Lastly, inspiring vignettes demonstrate the power of individual and collective actions that advance planetary health.

Nutrition Security for Planetary Health

Irana Hawkins, PhD, MPH, RDN

CRC Press
Taylor & Francis Group
Boca Raton London New York

CRC Press is an imprint of the
Taylor & Francis Group, an **informa** business

First edition published 2026
by CRC Press
2385 NW Executive Center Drive, Suite 320, Boca Raton FL 33431

and by CRC Press
4 Park Square, Milton Park, Abingdon, Oxon, OX14 4RN

CRC Press is an imprint of Taylor & Francis Group, LLC

ISBN: 9780367721206 (hbk)
ISBN: 9781003153528 (ebk)

DOI: 10.1201/9781003153528

Typeset in Times
by Newgen Publishing UK

Dedication:

Quinto Inuma Alvarado
Kichwa Tribal Leader, Peruvian Rainforest

Berta Cáceres
Lenca Woman and Co-founder,
National Council of Popular and Indigenous Organizations of Honduras

Homero Gómez González
Director, El Rosario Sanctuary, Ocampo, Michoacán, Mexico
Monarch Butterfly Biosphere Reserve | UNESCO World Heritage Site

Anton Mzimba
Head Ranger, Timbavati Private Nature Reserve, Greater Kruger National Park
UNESCO Biosphere Reserve, Kruger to Canyon, South Africa

Raúl Hernández Romero
Tour Guide, El Rosario Sanctuary, Ocampo, Michoacán, Mexico
Monarch Butterfly Biosphere Reserve | UNESCO World Heritage Site

…and all other wildlife defenders who were murdered as they worked to protect the natural environment.

Contents

Preface

"Everything starts now".

Miranda July[1], Artist, Author, Filmmaker

I've had a keen awareness of the grandeur of the natural environment and training in resource conservation since childhood in my hometown of Laguna Beach, a small coastal city in southern California (United States of America). Residents worked to protect the integrity of the natural environment, while conserving resources was mandatory in our home. So was the exploration of natural wonders. My father took us kids in our 1973 Ford Econoline camper van to witness the awe-inspiring splendors of the national parks including the mighty Redwoods and Sequoias, the Painted Desert, Death Valley, the Grand Canyon, Bryce Canyon, and Zion National Park. Fortunately, I was able to see these spectacular places twice due to the age span of my siblings. These experiences likely cemented my bond to the natural environment that became increasingly powerful over time.

When gazing at the ocean horizon as a child, I was not fully aware of the extent that nature offered *life*—or the perils biodiversity could face. Whales and phytoplankton are associated with every other breath that humans take yet the southern resident Killer whales (Orcas) of the Salish Sea (state of Washington) are endangered along with several other species of whales as indicated by the Washington Department of Fish & Wildlife. The ocean is acidifying and the splendor of life underwater has experienced unfathomable abuses. Today, the mass extinction of biodiversity is a reality. The systems humans have constructed are both widespread and destructive with the most vulnerable (humans and non-humans alike) enduring the greatest hardship. Furthermore, the World Meteorological Association reports that we likely surpassed our warming threshold of 1.5° Celsius in 2024.

Using the lens of an experienced Registered Dietitian Nutritionist (RDN) with extensive interdisciplinary training, this book delineates the magnitude of the problems we face. Moreover, *Nutrition Security for Planetary Health* offers solutions that can improve human health and bolster the natural processes of the Earth system which are unraveling before our very eyes. As the novel coronavirus COVID-19 (SARS CoV-2) offered yet another wake-up call to create a different way of life that is just for all living beings moving forward—we must engage in the radical act of walking the talk and create healthier outcomes for all. Do we truly comprehend that when the Intergovernmental Panel on Climate Change declared that we must decrease greenhouse gas emissions by 45% of 2010 levels by 2030 *that we have less than a decade to cut emissions in half*?

As a credentialed healthcare provider whose code of ethics mandates protecting public health, it is a matter of duty to reverse injustices, mitigate the burden of disease, and to protect ecosystems. Ultimately, the evidenced-based data are clear—we have no time to waste. It is my hope that more people and communities become inspired to drive improvements for planetary health that I write this book.

By improving planetary health with nutrition, nature, and equity, we simultaneously engage numerous co-benefits that create justice for the exquisite bounty of living beings and living systems of the planet. Indeed, what we put on our plates to healthfully nourish our bodies along with healthy lifestyle behaviors are far-reaching tools to nourish the planet, too. With billions of people on the planet, there are billions of solutions to enact daily.

NOTE

1 Paper Magazine. (2015, October 15). *Miranda July "Everything starts now"*. [Post]. X. https://x.com/4Prina/status/658340855275737088?mx=2

About the Author

Irana Hawkins, PhD, MPH, RDN, is a planetary health specialist and a scholar-practitioner. Her singular contributions have spanned clinical nutrition and public health, serving people from all walks of life; higher education including teaching and mentoring at the doctoral level; and planetary health research. Dr. Hawkins is a Certified California Naturalist, a Native Plant Steward, and a Master Recycler-Composter. She has first-authored dozens of publications and edited *Promoting Biodiversity in Food Systems* (CRC Press, 2019).

Acknowledgements

This book was written in a variety of ecoregions along the west coast of the continental United States which are the ancestral lands of Indigenous persons including the Duwamish, Salish, Suquamish, Kalapuyan, Tongva, Acjachemen, and Yuhaaviatam. Luhui Whitebear, a member of Oregon's Coastal Band Chumash Nation, describes such acknowledgements as a way to demonstrate that, "...we care about our shared histories and present actions so to create a better future" (Oregon State University, 2025). Additionally, we must also acknowledge the plight of Indigenous biodiversity and our living systems that have been destroyed or degraded, too.

The evidenced-based data that informs this book is both shocking and inspiring. We owe a debt of gratitude to those who work to advance the common good, whether it be through the scientific method or traditional ecological knowledge.

Lastly, I am most grateful to Ms. Randy Brehm, Senior Editor of Chemistry and Life Sciences at CRC Press and her team for the opportunity to illuminate the problems we face that *we can* resolve together.

REFERENCE

Oregon State University. (2025). *Land acknowledgement.* Institutional Diversity. https://diversity.oregonstate.edu/feature-story/land-acknowledgement

SECTION ONE

An Epic Array of Interconnected Crises

Far beyond the Carbon Emissions of Your Plate

1

"It really boils down to this: that all life is interrelated. We are all caught in an inescapable network of mutuality, tied into a single garment of destiny. Whatever affects one destiny, affects all indirectly."

Dr. Martin Luther King, Jr.[1], Christmas Eve, 1967

OVERVIEW

Over the past three decades, the rates of nutrition-related chronic disease have skyrocketed nationally and globally; the degradation of the natural environment is far-reaching; the chasm of wealth, poverty, and injustices has widened instead of narrowed; and "business-as-usual" practices increasingly derail the health of people, biodiversity, and the planet. As a result of these grave circumstances, moving forward with the interconnected and imperative constructs of nutrition, nature, and equity to improve planetary health is a timely solution for our multifaceted environmental and humanitarian crises. More specifically, *Nutrition Security for Planetary Health* informs the urgency of immediate actions with the power of one's plate and lifestyle behaviors; prioritizing and bolstering nature; and creating equity for the disenfranchised. Without rectifying the problems of these seemingly disparate yet intertwined circumstances, we face unprecedented challenges within the Earth System. This first chapter will introduce the reader to the major concepts of this book and starts by reflecting on our recent history marked by the novel coronavirus

DOI: 10.1201/9781003153528-2

COVID-19 (SARS-CoV-2) pandemic which underscored structural injustices where nature is included amongst the downtrodden.

NATURE, THE NOVEL CORONAVIRUS PANDEMIC, AND INJUSTICES

The year 2020 started with a series of events that accentuated egregious problems residing in plain sight. Scientists declared 2020 "Nature Super Year" where humanity was to boldly support nature due to the fragile state of biodiversity and the Earth System (United Nations, 2020). However, Nature Super Year was upended by COVID-19 and when sheltered-in-place, the world witnessed the extrajudicial killing of Mr. George Floyd that sparked outrage and protest (Kirka, 2020). Despite data indicating that some improvements were bestowed upon the natural environment when global travel came to a halt, we either witnessed or endured unprecedented climate-related disasters including wildfire, heat waves, tornadoes, hurricanes, and rising carbon dioxide emissions (Braga et al., 2020; World Meteorological Organization, 2021). We indeed are living in the clutches of injustices including climate change where scientists warn that without immediate action, our livelihoods will become increasingly challenging if not impossible for many (Stockholm Resilience Centre, 2023). Ending environmental degradation, bolstering biodiversity, and reversing injustices coincides with the United Nation's Sustainable Development Goals too as depicted in Figure 1.1 (United Nations, 2024a).

While the impact of individual behaviors may be downplayed and while we may not be able to change destructive policies tomorrow—we can align our individual values and our collective actions to support planetary health *today*. Despite the cacophony of attacks that can easily spread via social media and can be shared instantly—we can use our personal power to create positive outcomes (Garcia et al., 2019). Ergo, this book champions that essential solutions are within reach—and can jumpstart rebuilding our constructed systems. Indeed, we can nourish the entirety of life on planet Earth by improving planetary health.

FIGURE 1.1 Sustainable Development Goals, United Nations, 2024.

COMPREHENSIVE ACTIONS FOR PLANETARY HEALTH

While debates have raged over the years regarding the merits of a single action or an ideology to improve environmental outcomes—the dire state of the Earth System begs that we rapidly enact all solutions that eliminate damage to the natural environment and expediently mitigate the planetary boundaries that have been breached (Project Drawdown, 2020). Along with the importance of what we eat, this book aligns with key principles:

1) We must immediately engage every pro-environmental and nature-positive behavior with integrity and in perpetuity at all levels (individual, community, regional, global, policy, etc.)
2) Our individual actions do make an important difference; we can no longer view our personal actions in isolation and must comprehend the interconnected and far-reaching consequences of our lifestyle behaviors including food choices that impact the living systems of the planet
3) One positive action can create numerous and simultaneous co-benefits
4) Small actions over large numbers of people can make a big difference
5) There is no one solution or one approach to the dire multifaceted ecological crises we face—but a multitude of solutions where millions of people including the business community and policy makers can expedite beneficial outcomes
6) Many solutions to the ecological crises are cost-efficient and may require little to no technological inputs—and can increase positive humanistic interactions, too
7) Equity for the marginalized including nature must be a foundational priority as we reinvent and realign human-constructed systems.

WHOLE PLANT FOODS NUTRITION FOR PLANETARY HEALTH

The term whole plant foods nutrition (WPFN) is specifically used in this book to draw attention to the transformative action of consuming unprocessed whole plant foods and the *nutrition* they embody as a foundational dietary pattern to maximize nutrient intake while maximizing the health of the natural environment (Poore & Nemeck, 2018; Scarborough et al., 2023; Springmann et al., 2024). The term planetary health encompasses the inextricable relationship between human health and the health of the natural environment where degrading the natural environment degrades human health and vice versa where ultimately, the goal is healthy people living on a healthy planet (Rockefeller Foundation, 2017). Importantly, cultures across the globe have long embraced whole plant foods.

However, choosing whole plant foods nutrition as a tasty and high-quality dietary pattern is an underutilized solution for planetary health in high-income countries (Springmann et al., 2018). Yet, despite gains in the popular culture and calls from intergovernmental scientists to shift dietary patterns away from animal products to mitigate climate change—we are not seeing a broad-scale transition to whole plant foods nutrition. Dietary patterns high in animal products, fats, sugars, and processed foods are directly linked to chronic disease morbidity and mortality while inequities exacerbate this burden of disease, characterized by the term syndemic (Lancet, 2017). A syndemic describes diseases or health conditions that are exacerbated by inequities, multiplying the magnitude of the injustices faced (Lancet, 2017).

Plant foods can be grown to nourish the natural environment using agroecology and/or regenerative methods—or they can degrade the natural environment as is the case with conventional/industrial agriculture (Hashemi et al., 2024). However, some of the plant foods we routinely consume are destructive to the natural environment and biodiversity such as palm oil which is ubiquitous in food and household products (Gatti et al., 2019; Gatti & Velichevskaya, 2020). Bananas that are grown using fungicides can impair ecosystems and biodiversity while impacting the health of humans as well (Brühl et al., 2023; Grant et al., 2013). Transforming midwestern U.S. native grasslands to soy and corn monoculture crops has degraded soils, releasing carbon into the atmosphere—and destroyed biodiversity (Zhang et al., 2021). Hence, we cannot promote a diet for planetary health without recognizing all negative externalities (Wyckhuys et al., 2020). The accessibility of healthy whole plant foods is not a reality for many Americans though, while overall Americans still fall short of consuming the recommended servings of fruits and vegetables, legumes, nuts and seeds daily (U.S. Department of Agriculture & U.S. Department of Health and Human Services, 2020; Bozeman et al., 2020).

The commonly used term "whole foods plant-based diet" (WFPBD) has varying definitions such as dietary patterns that may or may not include animal products. Some plant-based foods are processed or ultra-processed and are associated with negative health outcomes (Satija et al., 2017). Many plant-food products are wrapped in plastic packaging that is known to be detrimental to the natural environment and biodiversity—with increasing evidence of concern for the human body, too (Landrigan et al., 2023). Hence, the term "whole plant foods nutrition" appears most suitable given the multifaceted planetary crises.

PRIVILEGE AND PLANETARY HEALTH

In addition to the premise that we can improve planetary health with whole plant foods nutrition, bolstering nature, and creating equity—the data is clear that those of privilege substantially contribute to environmental degradation (Cozzi et al., 2023). While this topic is further addressed in Chapter 2, it's long acknowledged that the most vulnerable humans have contributed the least to climate change (United Nations Development Programme [UNDP], 2023). Nature itself is included amongst the downtrodden as Pope Francis reminds us in the Encyclical Letter *On Care for Our Common Home* (UNDP, 2023; Pope Francis, 2015). While this book cannot unpack the root causes or the magnitude of all injustices, systems, and economic practices that wreak havoc on planetary health, *Nutrition Security for Planetary Health* will discern that we are indeed powerful in our ability to enact solutions.

CREATING JUSTICE FOR ALL

The third Monday in January is the designated federal holiday in the U.S. where we recognize the life's work of Dr. Martin Luther King, Jr., the indefatigable civil rights champion, Nobel Peace Laureate, and advocate of creating "The Beloved Community" (The King Center, n.d.). When accepting the Nobel Prize on behalf of all of people working to end injustices, he recognized that people everywhere have the right to nourishing food, education, dignity, equality, and freedom (King, 1964). The charge today is to produce nutritious foods that are continuously available to all people (nutrition security) while simultaneously supporting every aspect of planetary health.

While Dr. King's vision was considered radical, his message rings true today where a radically different pathway is the best route forward where changes in the constructed culture and global policies are desperately needed. As a global community though, we struggle to meet the standards we've set (United Nations, 2024b).

As a college student proclaimed at a 2019 climate rally in Eugene, Oregon (U.S.), "…as a destructive species, we humans have the opportunity to turn around this strange situation we've created and forge a positive future for all" (Hawkins, 2019). Truly, this is our opportunity to stand up and transcend our environmental and humanitarian failures by creating the best outcomes for planetary health—where whole plant foods nutrition, bolstering nature, and creating equity are foundational solutions.

NOTE

1 King, Jr., M.L. (1967, December 24). *Christmas eve sermon on peace and non-violence.* www.youtube.com/watch?v=1jeyIAH3bUI

REFERENCES

Bozeman, J.F., Springfield, S., & Theis, T.L. (2020). Meeting EAT-Lancet food consumption, nutritional, and environmental health standards: A U.S. case study across racial and ethnic subgroups. *Environmental Justice*, *13*(5), 160–172.

Braga, F., Scarpaa, F.M., Brando, V.E., Manfèa, G., & Zaggia, L. (2020). COVID-19 lockdown measures reveal human impact on water transparency in the Venice Lagoon. *Science of the Total Environment*, *736*, 1–7. https://doi.org/10.1016/j.scitotenv.2020.139612

Brühl, C.A., Andres, M.A., Echeverría-S´aenz, S., Bundschuh, M., Knabel, A., Mena, F., Petschick, L.L., Ruepert, C., & Stehle, S. (2023). Pesticide use in banana plantations in Costa Rica – A review of environmental and human exposure, effects and potential risks. *Environment International*, *174*, 1–13. https://doi.org/10.1016/j.envint.2023.107877

Cozzi, L., Chen, O., & Kim, H. (2023, February 22). The world's top 1% of emitters produce over 1000 times more CO2 than the bottom 1%. *International Energy Agency.* www.iea.org/commentaries/the-world-s-top-1-of-emitters-produce-over-1000-times-more-co2-than-the-bottom-1

Garcia, D., Galaz, V., & Daume, S. (2019). EATLancet vs yes2meat: The digital backlash to the planetary health diet. *The Lancet*, *394*, 2153–2154.

Gatti, R.C., & Velichevskaya, A. (2020). Certified "sustainable" palm oil took the place of endangered Bornean and Sumatran large mammals habitat and tropical forests in the last 30 years. *Science of the Total Environment*, *742*, 1–11. https://doi.org/10.1016/j.scitotenv.2020.140712

Gatti, R.C., Liang, J., Velichevskaya, A., & Zhou, M. (2019). Sustainable palm oil may not be so sustainable. *Science of the Total Environment*, *652*, 48–51. https://doi.org/10.1016/j.scitotenv.2018.10.222

Grant, P.B.C., Woudneh, M.B., & Ross, P.S. (2013). Pesticides in blood from spectacled caiman (*Caiman Crocodilus*) downstream of banana plantations in Costa Rica. *Environmental Toxicology and Chemistry*, *32*(11), 2576–2583.

Hashemi, F., Mogensen, L., van der Werf, H.M.G., Cederberg, C., & Knudsen, M.T. (2024). Organic food has lower environmental impacts per area unit and similar climate impacts per mass unit compared to conventional. *Communications Earth & Environment*, 5(250), 1–18. https://doi.org/10.1038/s43247-024-01415-6

Hawkins, I. (2019, October 1). I'm with the kids!. *The Register-Guard*. www.registerguard.com/story/opinion/letters/2019/10/02/i-m-with-kids/2638170007/

King, M.L.K. (1964, December 10). *Martin Luther King Jr.–acceptance speech*. The Nobel Prize. www.nobelprize.org/prizes/peace/1964/king/acceptance-speech/

Kirka, D. (2020, May 31). *World alarmed by violence in US; thousands march in London*. The Associated Press. www.yahoo.com/news/world-uneasily-watches-us-protests-130003544.html

Lancet [Editorial]. (2017). Syndemics: Health in context. *The Lancet*, *389*, 881.

Landrigan, P.J., Raps, H., Cropper, M., Bald, C., Brunner, M., Canonizado, E.M., Charles, D., Chiles, T.C., Donohue, M.J., Enck, J., Fenichel, P., Fleming, L.E., Ferrier-Pages, C., Fordham, R., Gozt, A., Griffin, C., Hahn, M.E., Haryanto, B., Hixson, R., Ianelli, H., & Dunlop, S. (2023). The Minderoo-Monaco commission on plastics and human health. *Annals of Global Health*, *89*(1), 1–215. https://doi.org/10.5334/aogh.4056

Poore, J., & Nemeck, T. (2018). Reducing food's environmental impacts through producers and consumers. *Science*, 360, 987–992.

Pope Francis. (2015). *Laudato Si'* [On Care for Our Common Home] [Encyclical letter]. Vatican Press. www.vatican.va/content/francesco/en/encyclicals/documents/papa-francesco_20150524_enciclica-laudato-si.html

Project Drawdown. (2020). *The drawdown review: Climate solutions for a new decade*. https://drawdown.org/sites/default/files/pdfs/TheDrawdownReview–2020–Download.pdf

Rockefeller Foundation. (2017). *Panorama perspectives on health: Conversations on planetary health*. www.rockefellerfoundation.org/wp-content/uploads/Planetary-Health-101-Information-and-Resources.pdf

Satija, A., Bhupathiraju, S.N., Spiegelman, D., Chiuve, S.E., Manson, J.E., Willett, W., Rexrode, K.M., Rimm, E.B., & Hu, F.B. (2017). Healthful and unhealthful plant-based diets and the risk of coronary heart disease in U.S. adults. *Journal of the American College of Cardiology*, *70*(4), 411–422. http://dx.doi.org/10.1016/j.jacc.2017.05.047

Scarborough, P., Clark, M., Cobiac, L., Papier, K., Knuppel, A., Lynch, J., Harrington, R., Key, T., & Springmann, M. (2023). Vegans, vegetarians, fish-eaters and meat-eaters in the UK show discrepant environmental impacts. *Nature Food*, *4*(7), 565–574. https://doi.org/10.1038/s43016-023-00795-w

Springmann, M. (2024). A multicriteria analysis of meat and milk alternatives from nutritional, health, environmental, and cost perspectives. *Proceedings of the National Academy of Sciences*, *121*(50), 1–9. https://doi.org/10.1073/pnas.2319010121

Springmann, M., Wiebe, K., Mason-D'Croz, D., Sulser, T.B., Rayner, M., & Scarborough, P. (2018). Health and nutritional aspects of sustainable diet strategies and their association with environmental impacts: A global modelling analysis with country-level detail. *The Lancet Planetary Health*, *2*, e451–e461.

Stockholm Resilience Centre. (2023). *Planetary boundaries*. www.stockholmresilience.org/research/planetary-boundaries.html

The King Center. (n.d.). *The King Center's definition of non-violence*. https://thekingcenter.org/about-tkc/the-king-philosophy/

U.S. Department of Agriculture & U.S. Department of Health and Human Services. (2020). *Dietary guidelines for Americans, 2020-2025. 9th edition*. US Department of Agriculture and US Department of Health and Human Services www.dietaryguidelines.gov/sites/default/files/2020-12/Dietary_Guidelines_for_Americans_2020-2025.pdf

United Nations. (2024a). *Sustainable development goals communications materials.* United Nations. www.un.org/sustainabledevelopment/news/communications-material/

United Nations. (2024b). The sustainable development goals report 2024. United Nations. https://unstats.un.org/sdgs/report/2024/The-Sustainable-Development-Goals-Report-2024.pdf

United Nations. (2020, March 21). Turn around deforestation in 2020, the 'Nature Super Year', says Guterres. *UN News.* https://news.un.org/en/story/2020/03/1059852#:~:text=On%20the%20…rnational%20Day%20of,on%20deforestation%20and%20forestry%20loss

United Nations Development Programme. (2023, June 30). *Climate change is a matter of justice–here's why.* United Nations.

World Meteorological Association. (2021). *State of the Global Climate* 2020. https://library.wmo.int/viewer/56247/download?file=1264_Statement_2020_en.pdf&type=pdf&navigator=1

Wyckhuys, K.A., Aebi, A., van Lexmond, M.F.B., Bojaca, C.R., Bonmatin, J.M., Furlan, L., ... & Ikenaka, Y. (2020). Resolving the twin human and environmental health hazards of a plant-based diet. *Environment International, 144*, 1–11. https://doi.org/10.1016/j.envint.2020.106081

Zhang, X., Lark, T.J., Clark, C.M., Yuan, Y., & LeDuc, S.D. (2021). Grassland-to-cropland conversion increased soil, nutrient, and carbon losses in the US Midwest between 2008 and 2016. *Environmental Research Letters, 16*, 1–13. https://doi.org/10.1088/1748-9326/abecbe

A Planet with Too Many in Peril

2

"Right now, the earth is full of refugees, human and not, without refuge."

Donna Haraway[1], PhD, Distinguished Professor Emerita, University of California Santa Cruz

OVERVIEW

The spectacular array of extraordinary living beings (human and non-human alike) and our living systems of the planet interact by way of natural cycles on land, in the ocean, with the atmosphere and magnetic poles, and by deep Earth processes; this is called the Earth System (International Geosphere-Biosphere Programme, n.d.). Biodiversity and the living systems of the natural environment underpin our lives on this planet (Intergovernmental Science-Policy Platform on Biodiversity and Ecosystem Services [IPBES], 2019). Yet despite the vast grandeur of life on this planet, the Earth System is being degraded and exploited by way of anthropogenic practices (human activity)—even though humans are comprised of and need the very elements we are destroying (IPBES, 2019). This chapter starts by framing the humanitarian inequities experienced around the globe and ends by summarizing the threats to the living systems of the planet. We cannot improve planetary health without acknowledging how our histories and hegemonic systems impact the entirety of the living beings on the planet.

DOI: 10.1201/9781003153528-3

THE DISPROPORTIONATE IMPACT OF THE COVID-19 PANDEMIC

Widespread global awareness of COVID-19 started in 2020 where this pathogen not only ended scores of lives and changed livelihoods throughout the world—but underscored glaring injustices (Centers for Disease Control and Prevention, 2023). As of January 7, 2025, over 7 million lives were lost around the globe (World Health Organization, 2024). The COVID-19 pandemic disproportionately impacted vulnerable communities throughout the world where some countries did not have sufficient monies or infrastructure to combat the disease (U.S. Global Leadership Coalition, 2021). The COVID-19 pandemic also accentuated the connections between environmental degradation and the food system that will be discussed in detail in Chapter Four—and highlighted leadership deficits, too (deSouza et al., 2022).

WAR, DISPLACEMENT, CLIMATE REFUGEES, AND FOOD INSECURITY

Over 100 countries across the globe are involved in armed conflict where an estimated one in six persons globally was exposed to violence in 2024 (Armed Conflict Location & Event Data [ACLED], 2024; Geneva Academy, n.d.). War destroys not only the living beings and the living systems of the planet—but uses massive amounts of fossil fuels that exacerbates the destruction (Crawford, 2019; Hryhorczyk et al., 2024). Globally, the number of forcibly displaced persons continues to increase; by May 2024, 120 million people were forced to relocate due to war, conflict, climate disasters, and human rights violations where approximately two-thirds of these persons live in poverty (European Commission, 2024). Climate change impacts the most vulnerable where globally, nearly 90% of climate refugees are from the most vulnerable countries where Black, Indigenous, and Persons of Color (BIPOC) are disproportionately represented (Internal Displacement Monitoring Center, 2024; United Nations [UN], 2021).

Importantly, a recent analysis quantifying the human cost of global warming demonstrated that during 1960–1990, over 600 million people fell outside of the "human climate niche"—the regions of the world where climatic conditions can support human life (Lenton, 2023). As the lead author of this research study stated, "Those people who are affected are the poorer people on the planet. At higher temperatures, life becomes unbearable, affecting water, agriculture, and food." (Silverstein, 2023). Without a livable climate, the risk of morbidity and mortality increases as well as stressors such as food insecurity as large numbers of people in vulnerable regions depend on local farming (Lenton, 2023). If global warming reaches 2.7° Celsius by the year 2070, 1,997 million people in 112 countries will endure unprecedented heat where India, Nigeria, Indonesia, Philippines, Pakistan, Sudan, Niger, Thailand, Saudi Arabia, and Burkina Faso will be heavily impacted (Lenton, 2023). By the end of the century, it's projected that approximately one-third of the global population will live outside the human climate niche (Lenton, 2023).

Displaced persons face increased food insecurity (UN High Commissioner for Refugees, 2021). Despite the 1948 Universal Declaration on the Right to Food, more than one-third of the global population could not afford a healthy diet in 2022 where low-income countries are disproportionately impacted (UN High Commissioner on Human Rights, n.d.; Food and Agriculture Organization [FAO] et al., 2024). According to Secretary-General Antonio Guterres, "Hunger is no longer about lack of food. It is largely a man-made disaster—concentrated in countries affected by large-scale, protracted conflict" (Guterres, 2021).

POVERTY & INJUSTICES

Globally, over one billion people live in slum-like conditions while 40% do not have access to safe drinking water (IPBES, 2019; UN, 2025). More than one in five across the globe (1.3 billion people) live in multidimensional poverty where half are under the age of 18 (Alkire et al., 2024). Multidimensional poverty encompasses factors including but not limited to nutrition, health, child mortality, school attendance and years of schooling, sanitation, housing, lack of electricity, lack of access to improved drinking water, etc. (Alkire et al., 2024). Nearly 50% live in Sub-Saharan Africa while one-third live in South Asia (Alkire et al., 2024). India, Pakistan, Ethiopia, Nigeria, and the Democratic Republic of the Congo comprise nearly half of the 1.1 billion persons living in poverty (Alkire et al., 2024). Additionally, half of the global population struggles to meet basic needs (World Bank Group, 2020). Those that are poor earn less than 25% of income globally as wealth is increasingly concentrated amongst the top 1% which reinforces intergenerational poverty and inequalities (United Nations Department of Economic and Social Affairs, 2020). A report assessing family income, poverty, and mobility in the U.S. found that over 20% of Black Americans compared to just over 1% of White Americans experience poverty spanning over three generations (Winship et al., 2021).

PLUNDERING THE EARTH SYSTEM

Transgressing our Planetary Boundaries

The Planetary Boundaries Concept is a framework that conceptualizes the overall stability and interconnectedness of the Earth System based on the totality of the collective impact (Steffen et al., 2015). It encapsulates nine key systems represented in **Figure 2.1** (Richardson, et al., 2023). Moving away from our "safe operating space" at the center of the diagram towards the outer circumference delineates a very high risk to humanity.

Six planetary of nine boundaries have now been transgressed: climate change, land-system change, biosphere integrity, biogeochemical flows, freshwater use, and novel entities as indicated in Figure 2.1 (Richardson, et al., 2023). We've been moving in the wrong direction for decades as depicted in Figure 2.2.

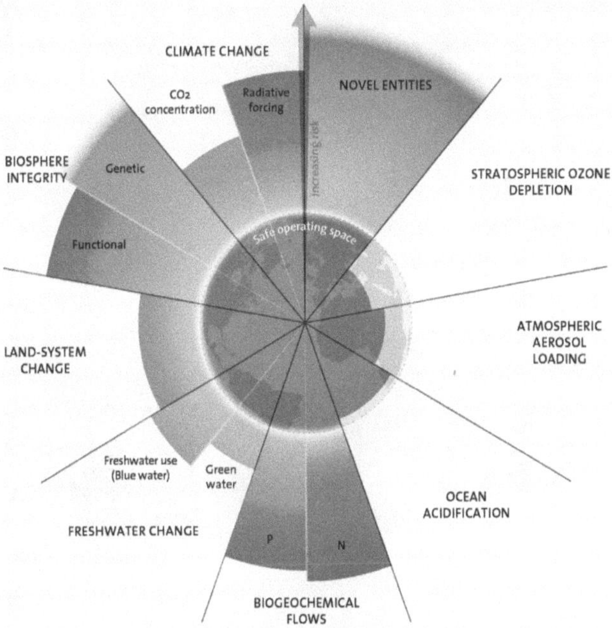

FIGURE 2.1 Update to the Planetary Boundaries, Azote for the Stockholm Resilience Centre, 2023.

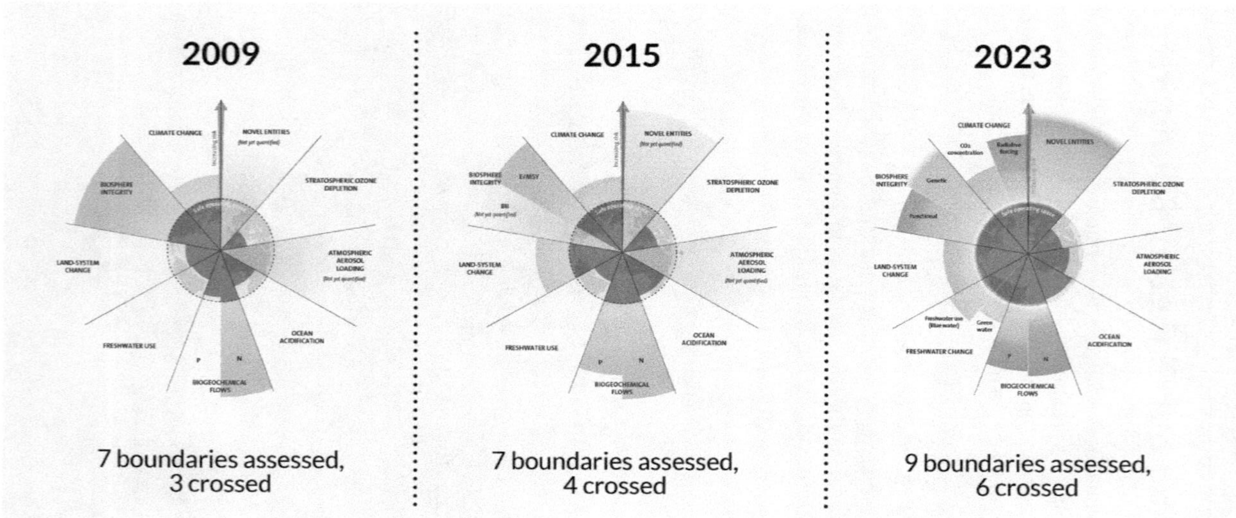

FIGURE 2.2 The Evolution of the Planetary Boundaries Framework, Azote for Stockholm Resilience Centre, 2023.

Novel entities have been transgressed by way of chemical and plastics pollution (Persson et al., 2022). Leading scientists believe that the planetary boundary for ocean acidification has *likely* been transgressed, too (Steffen & Morgan, 2021). These transgressions point to the dire need for meaningful change across all sectors and all facets of the Earth System—both globally and locally. Importantly, the food system is implicated in all transgressions of the planetary boundaries. Additionally, some scientists suggest redefining the Planetary Boundaries to address a myriad of critical needs including social equity and local metrics—while imploring bottom-up actions at all levels (Downing et al., 2019; Li et al., 2021).

Climate Change and Greenhouse Gas Emissions (GHGE)

Climate change can be defined as long-term changes to weather patterns, primarily due to human activity that has resulted in generating heat-trapping greenhouse gases such as carbon dioxide, methane, and nitrous oxide (National Aeronautics Space Administration [NASA], 2022). The Intergovernmental Panel on Climate Change (IPCC) reports that, "Human influence has warmed the climate at a rate that is unprecedented in at least the last 2000 years" (Intergovernmental Panel on Climate Change [IPCC], 2021). Climate scientist Friedericke Otto conveyed that record-setting heat, "It's a death sentence for people and ecosystems" (Dickie, 2023). While the planet is warming, some areas have become colder—and weather extremes are prevalent. The U.S. experienced another year of billion-dollar climate disasters in 2024 that included drought, flooding, severe storm events, tropical cyclones, wildfire, and winter storms—27 in total that are attributed to climate change (National Oceanic Atmospheric Administration [NOAA] National Centers for Environmental Information [NCEI], 2025).

Decades ago, scientists declared that if we are to maintain the living systems of the Earth System, we must reduce global carbon emissions to a range of 300-350 ppm of carbon dioxide to avoid deleterious outcomes which included recommendations such as transforming agricultural and forestry practices (Hansen et al., 2008). The goal now is to avert a "hothouse Earth trajectory" that is driven by continued anthropogenic biosphere degradation and rising greenhouse gas emissions (Steffen et al., 2018). As of November 2024, global carbon dioxide (CO_2), emissions were measured at 424 ppm—well over scientific recommendations (NASA, 2025). Despite the COVID-19 pandemic that halted global travel and altered the production and flow of goods and services, the year 2021 was one of the warmest years on record (World Meteorological Organization, 2021). In fact, 2021 was a record-breaking year for four key indicators of climate change—ocean acidification, ocean heat, sea level rise, and GHGE—amidst exceptional drought, heat waves, floods, displacement, food insecurity, and ecosystem degradation (World Meteorological Organization, 2021). These interrelated physical indicators demonstrate the interconnected and deleterious outcomes experienced globally that are linked to anthropogenic climate change.

Both methane (CH_4) and nitrous oxide (N_20) are powerful greenhouse gasses (World Meteorological Organization, 2022). Synthetic nitrogen inputs cause a myriad of problems including waterway degradation (Donner, 2006). Carbon dioxide (CO_2) levels are higher than any time over the last 2 million years and methane is higher than any time over the past 800,000 years, respectively (IPCC, 2021). Global mean sea levels are the highest in 3,000 years while the ocean is the warmest in 11,000 years (IPCC, 2021). If GHGE are not controlled, scientists project that the land and the ocean will continue to rise in temperature, exacerbating atmospheric GHGE and altering feedback loops, creating irreversible damage (IPCC, 2021; Steffen et al., 2018).

Wildfire, Heatwaves, and Temperature Extremes

Climate change, heat extremes, and drought increases the risk of wildfire (United Nations Environment Programme [UNEP], 2022). Wildfire is associated with a myriad of negative outcomes such as the loss of biodiversity and ecosystem services; the loss of human life; changes to the carbon cycle, altering feedback loops; creating pollution and toxicants; and massive economic losses (UNEP, 2022). Globally, wildfire is projected to increase while in 2021, wildfire ravaged countries across the globe including but not limited to California (USA), Algeria, Greece, North Macedonia, Turkey, France, Italy, Tunisia, Morocco, Serbia, Lebanon, Israel, and Libya (UNEP, 2022; World Meteorological Association, 2022). The catastrophic 2019–2020 fires in Australia killed or injured approximately 3 billion animals where animals are indispensable in the Earth System (van Eeden et al., 2020).

The "Triple Threat"

Global environmental scientists are now calling the multifaceted climate crises including climate change, biodiversity loss, and pollution "the triple threat" (United Nations Framework Convention on Climate Change, 2022). In 2019, air pollution claimed over 6.5 million lives globally; water pollution 1.4 million lives; and lead

(Pb) claimed 900,000 lives, respectively (Fuller et al., 2022). Whether stemming from industry or agriculture, pollution is degrading our land, air, and waterways—and is responsible for one in six human deaths, globally (Fuller et al., 2022). Over 90% of these deaths occur in low-and-middle income countries (Fuller et al., 2022). Additionally, integrated assessment models in the U.S. have demonstrated that food-animal production (terrestrial animals and aquaculture) surpasses electricity suppliers as the largest contributor to air pollution and fine particulate matter where the environmental damage of food-animal production surpasses the value added (Tschofen et al., 2019).

Pollution is not confined to a particular region per se and may be distributed by way of water, air, the food system, consumer products, etc. (Fuller et al., 2022). "Forever chemicals" is the moniker given to a group of human-made chemicals (per- and polyfluoroalkyl substances) that persist in the natural environment, accumulating in waterways, the human body, and even food (Agency for Toxic Substances and Disease Registry [ASTDR], 2022). In 2022, the Michigan Department of Agriculture and Rural Development (U.S.) found that locally sold beef was contaminated with perfluorooctane sulfonic acid (Klase & Vig, 2022).

Pharmaceutical Pollution

Over 50% of U.S. rivers, streams, and creeks and 55% of lakes, ponds, and reservoirs along with 26% of bays, harbors, and estuaries are impaired for at least one important ecosystem service such as fish safe for human consumption or potable water (Kelderman et al., 2022). Pharmaceutical pollution is concerning as byproducts can accumulate in waterways and wildlife. In fact, when analyzing over 1,000 samples from rivers located in 104 countries and on every continent, researchers found 61 active pharmaceutical ingredients (Wilkinson et al., 2022). Metformin, a drug used for diabetes, was found amongst the highest concentrations (Wilkinson et al., 2022). In an assessment of 59 wadable streams in the Piedmont Region of the U.S. (Atlanta, Charlotte, Raleigh, and Washington, D.C.), metformin was pervasive (Bradely et al., 2016). With the exception of the Antarctica, the anti-hypertensive drug atenolol was found across all continents (Bradely et al., 2016). With six out of ten American adults plagued with a chronic disease and where four in ten have two or more, pharmaceutical pollution is alarming (Centers for Disease Control and Prevention, 2022). Notably, nutrition-related chronic diseases *can* be prevented with whole plant foods nutrition and lifestyle medicine which will be delineated throughout this book (Hawkins, 2019).

Mass Extinction

The rate of extinction of biodiversity has far exceeded background rates where scientists have called out a modern-day "Sixth Mass Extinction" (Ceballos et al., 2017). While extinctions are considered a normal part of evolutionary processes, contemporary extinction rates are neither normal or inevitable (Ceballos et al., 2017; Perfecto et al., 2009). The anthropogenic degradation of the planet has transformed and stressed biological systems (Barnosky et al., 2012).

The Intergovernmental Panel on Biodiversity and Ecosystem Services (IPBES) is an independent collective amongst international scientists; their 2019 landmark report assessed over 15,000 scientific publications globally where 85% of wetlands have been lost; three-quarters of all land has been substantially altered and degraded; two-thirds of the ocean is experiencing multiple impacts due to overfishing and over-exploitation that includes plastics pollution where only 3% of ocean is thought to be free from human pressures—and half of coral reefs have been lost (Intergovernmental Panel on Biodiversity and Ecosystem Services [IPBES], 2019). Biodiversity loss undermines the resilience and health of ecosystems (IPBES, 2019). Marine plastics pollution has negatively impacted 86% of turtles, 44% of seabirds, and 43% of marine mammals (IPBES, 2019). Plastics pollution will be discerned in forthcoming chapters, too.

Globally, nearly a million species are threatened with extinction while one-half million species are at risk due to insufficient habitat (IPBES, 2019). Since the 1500's, humans have driven 680 vertebrates to extinction (IPBES, 2019). The decline of our natural systems is extraordinary where the integrity of biological communities has declined and biomass has decreased by 82% (IPBES, 2019).

The International Union for Conservation of Nature (IUCN) is a collaboration between government entities, member organizations, and experts that work to assess and safeguard nature (International Union for Conservation of Nature [IUCN], n.d.). The IUCN Red List is an account of the well-being of biodiversity as well as their risk of extinction (IUCN, 2022). Currently, over 166,000 species globally are on the IUCN Red List where over 46,000 species threatened with extinction including, "…44% of reef building corals, 41% of amphibians, 38% of trees, 37% of sharks and rays, 34% of conifers, 26% of mammals, 26% of freshwater fishes and 12% of birds" (IUCN, 2025).

Noting that, "…dangerous tipping points are approaching", *The Living Planet Report 2024* documented an average 73% loss of monitored global biodiversity between the years 1970–2020 (World Wildlife Fund for Nature [WWF], 2024). Unsustainable and destructive agricultural practices and climate change are amongst the top drivers while overfishing drives ocean decline

(WWF, 2024). Overall, species loss is driven by agriculture, overexploitation, climate change, pollution, invasive species, and disease (WWF, 2024). Eighty-five percent (85%) of freshwater populations have plummeted while terrestrial populations have dropped by 69%; marine populations have declined by 56% (WWF, 2024). Lastly, to stress the importance of biodiversity in our daily food choices, three-quarters of the world's crops depend on pollination—and one in every three bites is attributed to a pollinator where over 200,000 species provide pollination services; losing pollinators increases the risk of food insecurity (FAO, 2025).

Soil Pollution

Soil is a precious resource that harbors an abundance of biodiversity where healthy soils perform a vast array of ecosystem services including: nutrient cycling; carbon sequestration; the degradation of contaminants; water sequestration, filtration, and management; supporting the health of plants and animals; and controlling pathogens and pests, etc. (Geisen et al., 2019). Scientists characterize soil health by way of productivity, diversity, resilience, and adaptation (Geisen et al., 2019). Soil takes a long time to build—but only a short amount of time to destroy. With degraded and denuded soils across the globe, restoring soil health is a critical priority (FAO & United National Environment Programme [UNEP], 2021). Importantly, 95% of the foods we consume are derived from soil (FAO & UNEP, 2021).

Soil pollution is a top driver of soil loss where soil is polluted by way of chemicals (organic and inorganic), agriculture, and industrial activities (FAO & UNEP, 2021). Globally, about 35 billion tons of soil are lost every year with one-third due to agriculture and agrichemicals (FAO & UNEP, 2021). Contaminants can also include pharmaceuticals, personal care products, plastics, phthalates & plasticizers, nanomaterials, and antimicrobials/antibiotics (FAO & UNEP, 2021). Soil pollution destroys soil biodiversity, hampers soil function, and can contaminate groundwater and waterways including creating dead zones (FAO & UNEP, 2021). Polluting and degrading soils can cause soil acidification, too (FAO & UNEP, 2021).

Light Pollution

Artificial light is an overlooked environmental pollutant that interferes with natural processes, creating negative ramifications for both humans and the living systems of the planet (DarkSky International, 2025). As an example, trees impacted by light pollution have an irregular loss of leaves while evening pollinators can experience negative outcomes, too (DarkSky International, 2025). Globally, light pollution is increasing (Kyba et al., 2017). Light pollution will be further discussed in Chapter 11. In closing, the anthropogenic degradation of the natural environment and the level of human suffering throughout the world is immense, interconnected, and multifaceted.

NOTE

1 Haraway, D. (2015). Anthropocene, Capitalocene, Plantationocene, Chthulucene: Making kin. *Environmental Humanities, 6*, 159–165.

REFERENCES

Agency for Toxic Substances and Disease Registry [ASTDR]. (2022, November 1). *What are PFAS?* Centers for Disease Control and Prevention. www.atsdr.cdc.gov/pfas/health-effects/overview.html

Alkire, S., Kanagaratnam, U., & Suppa, N. (2024). *A methodological note on the global Multidimensional Poverty Index (MPI) 2024 changes over time results for 86 countries.* Oxford Poverty & Human Development Initiative. https://hdr.undp.org/system/files/documents/hdp-document/mpireport2024en.pdf

Armed Conflict Location & Event Data [ACLED]. (2024). *ACLED conflict index.* https://acleddata.com/conflict-index/

Barnosky, A.D., Hadly, E.A., Bascompte, J., Berlow, E.L., Brown, J.H., Fortelius, M., Getz, W.M., Harte, J., Hastings, A., Marquet, P.A., Martinez, N.D., Mooers, A., Roopnarine, P., Vermeij, G., Williams, J.W., Gillespie, R., Kitzes J., Marshall, C., Matzke, N., … & Smith, A.B. (2012). Approaching a state shift in Earth's biosphere. *Nature*, *486*, 52–58. https://doi.org/10.1038/nature11018

Bradley, P.M. (2016). Metformin and other pharmaceuticals widespread in wadeable streams of the Southeastern United States. *Environmental Science & Technology Letters*, *3*, 243–249. https://doi.org/10.1021/acs.estlett.6b00170

Ceballos, G., Ehrlich, P.R., & Dirzo, R. (2017). Biological annihilation via the ongoing sixth mass extinction signaled by vertebrate population losses and declines. *Proceedings of the National Academy of Sciences*, *114*(30), E6089–E6096. https://doi.org/10.1073/pnas.1704949114

Centers for Disease Control and Prevention. (2022). *Chronic diseases in America.* National Center for Chronic Disease Prevention and Health Promotion. www.cdc.gov/chronicdisease/resources/infographic/chronic-diseases.htm

Centers for Disease Control and Prevention. (2023). *CDC Museum COVID-19 Timeline.* www.cdc.gov/museum/timeline/covid19.html

Crawford, N.C. (2019). *Pentagon fuel use, climate change, and the costs of war.* Watson Institute, Brown University. https://watson.brown.edu/costsofwar/files/cow/imce/papers/Pentagon%20Fuel%20Use%2C%20Climate%20Change%20and%20the%20Costs%20of%20War%20Revised%20November%202019%20Crawford.pdf

DarkSky International. (2025). *Definitions of light pollution.* https://darksky.org/resources/what-is-light-pollution/

deSouza, L.E.P.F., Castro, M.C., Carmo, E.H., & Polidoro, M. (2022). The global failure offacing the pandemic. *Global Health Action, 15*, 1–4. https://doi.org/10.1080/16549716.2022.2124645

Dickie, G. (2023, July 4). World registers hottest day ever recorded on July 3. *Reuters.* www.reuters.com/world/world-registers-hottest-day-ever-recorded-july-3-2023-0704/#:~:text=July%204%20%2D%20Monday%2C%20July%203,heatwaves%20sizzled%20around%20the%20world.

Donner, S.D. (2006). Surf or turf: A shift from feed to food cultivation could reduce nutrient flux to the Gulf of Mexico. *Global Environmental Change, 17*, 105–113.

Downing, A.S., Bhowmik, A., Collste, D., Cornell, S.E., Donges, J., Fetzer, I., Häyhä, T., Hinton, J., Lade, S., & Mooij, W.M. (2019). Matching scope, purpose and uses of planetary boundaries science. *Environmental Research Letters, 14,* 1–11. https://doi.org/10.1088/1748-9326/ab22c9

European Commission. (2024, June 20). *Forced displacement.* https://civil-protection-humanitarian-aid.ec.europa.eu/what/humanitarian-aid/forced-displacement_en

Food and Agriculture Organization [FAO]. (2025). *Pollinators are the unsung heroes of our ecosystems.* United Nations. www.fao.org/pollination/en

Food and Agriculture Organization [FAO]. (2024). *The state of food security and nutrition in the world.* United Nations. https://openknowledge.fao.org/server/api/core/bitstreams/d5be2ffc-f191-411c-9fee-bb737411576d/content

Food and Agriculture Organization [FAO] & United National Environment Programme [UNEP]. (2021). *Global assessment of soil pollution – Summary for policy makers.* FAO. https://openknowledge.fao.org/server/api/core/bitstreams/8fa95d84-bb05-4c77-b02e-425f70ba6834/content

Geisen, S., Wall, D.H., & van der Putten, W.H. (2019). Challenges and opportunities for soil biodiversity in the Anthropocene. *Current Biology, 29*, R1036–R1044.

Geneva Academy. (n.d.). *Today's armed conflicts.* Geneva University. https://geneva-academy.ch/galleries/today-s-armed-conflicts

Guterres, A. (2021, December 19). *Hunger is no longer about lack of food.* [Tweet]. UN World Food Programme.

Fuller, R., Landrigan, P.J., Balakrishnan, K., Bathan, G., Bose-O'Reilly, S., Brauer, M., Caravanos, J, Chiles, T., Cohen, A., Corra, L., Cropper, M., Ferraro, G., Hanna, J., Hanrahan, D., Hu, H., David Hunter, D., Gloria Janata, G., Rachael Kupka, R., & Lanphear, B. (2022). Pollution and health: A progress update. *Lancet Planetary Health, 6*, e535–e547. https://doi.org/10.1016/S2542-5196(22)00090-0

Hansen, J. (2008). Target atmospheric CO2: Where should humanity aim? *The Open Atmospheric Science Journal, 2*, 217–231.

Hawkins, I.W. (Ed.) (2019). *Promoting biodiversity in food systems.* CRC Press. https://doi.org/10.1201/b22084

Hryhorczuk, D., Levy, B.S., Prodanchuk, M., Kravchuk, O., Bubalo, N., Hryhorczuk, A., & Erickson, T.B. (2024). The environmental health impacts of Russia's war on Ukraine. *Journal of Occupational Medicine and Toxicology*, *19*(1), 1–14.

Intergovernmental Science-Policy Platform on Biodiversity and Ecosystem Services (IPBES). (2019). *Summary for policymakers of the global assessment report on biodiversity and ecosystem services of the Intergovernmental Science-Policy Platform on Biodiversity and Ecosystem Services.* IPBES. https://doi.org/10.5281/zonodo.3553579

Internal Displacement Monitoring Center. (2024). *Global Report on internal displacement.* IDMC. https://api.internal-displacement.org/sites/default/files/publications/documents/IDMC-GRID-2024-Global-Report-on-Internal-Displacement.pdf

International Geosphere-Biosphere Programme. (n.d.). *Earth system definitions.* IGBP. www.igbp.net/globalchange/earthsystemdefinitions.4.d8b4c3c12bf3be638a80001040.html

Intergovernmental Panel on Climate Change [IPCC]. (2021). Summary for policymakers. In *Climate change 2021: The physical science basis. Contribution of working group I to the sixth assessment report of the intergovernmental panel on climate change.* https://doi.org/10.1017/9781009157896.001.

International Union for Conservation of Nature [IUCN]. (n.d.). *About IUCN.* IUCN. https://iucn.org/about-iucn

International Union for Conservation of Nature [IUCN]. (2025). *Background & history.* IUCN. www.iucnredlist.org/about/background-history

Kelderman, K., Phillips, A., Pelton, T., Schaeffer, E., MacGillis-Falcon, P., & Bernhardt, C. (2022). *The clean water act at 50: Promises half-kept at the half-century mark.* Environmental Integrity Project. https://environmentalintegrity.org/wp-content/uploads/2022/03/CWA-report-UPDATED-8.9.23.pdf

Klase, A. & Vij, P. (2022). *Health Consultation: Evaluation of PFOS in frozen wholesale beef from cattle exposed to PFOS through contaminated crops and soil due to biosolids applications at a farm in Livingston County, Michigan.* Michigan Department of Health and Human Services Division of Environmental Health. www.michigan.gov/mdhhs/-/media/Project/Websites/mdhhs/Safety-and-Injury-Prevention/Environmental-Health/Health-Assessments/Documents/Grostic-Farms-Beef-HC-12072022-FINALAPPROVED.pdf?rev=77550275ce9041f6a6dda9682b363c61&hash=7E2327185A5679F8F862DA03DDD10E5D

Kyba, C.C.M., Kuester,T., Sánchez de Miguel, A., Baugh, K., Jechow, A., Hölker,F., Bennie, J., Elvidge, C.D., Gaston, K.J., & Guanter, L. (2017). Artificially lit surface of Earth at night increasing in radiance and extent. *Science Advances*, *3*(e1701528), 1–8.

Lenton, T.M., Loriani, S., Dunn, R.R., Xu, C., Abrams, J.F., Ghadiali, A., Sakschewski, B., Zimm, C., Ebi, K.L., Svenning, J.C., & Scheffer, M. (2023). Quantifying the human cost of global warming. *Nature Sustainability*, *6*, 1237–1247. https://doi.org/10.1038/s41893-023-01132-6

Li, M., Wiedmann, T., Fang, K., & Hadijikakou, M. (2021). The role of planetary boundaries in assessing absolute environmental sustainability across scales. *Environment International*, *152*, 1–13. https://doi.org/10.1016/j.envint.2021.106475

National Aeronautics Space Administration [NASA]. (2022). *Overview: Weather, global warming, and climate change*. https://climate.nasa.gov/resources/global-warming-vs-climate-change/

National Aeronautics Space Administration [NASA]. (2025, January 14). *Carbon dioxide: Latest measurement November 2024*. https://climate.nasa.gov/vital-signs/carbon-dioxide/?intent=121

National Oceanic Atmospheric Administration [NOAA] National Centers for Environmental Information (NCEI). (2025). *U.S. billion-dollar weather and climate disasters*. www.ncei.noaa.gov/access/billions/

Perfecto, I., Vandermeer, J., & Wright, A. (2009). *Nature's matrix: Linking agriculture, conservation, and food sovereignty*. Earthscan.

Persson, L., Almroth, B.M.C, Collins, C.D., Cornell, S., de Wit, C.A., Diamond, M.L., Fantke, P., Hassellöv, M., MacLeod, M., Ryberg, M.W., Jørgensen, P.S., Villarrubia-Gómez, P., Wang, Z., & Hauschild, M.W. (2022). Outside the safe operating space of the planetary boundary for novel entities. *Environmental Science & Technology*, *56*, 1510–1521. https://doi.org/10.1021/acs.est.1c04158

Richardson, K., Steffen, W., Lucht, W., Bendtsen, J., Cornell, S.E., Donges, J.F., Drüke, M., Fetzer, I., Bala, G., von Bloh, W., Feulner, G., Fiedler, S., Gerten, D., Gleeson, T. Hofmann, M., Huiskamp, W., Kummu, M., Mohan, C., Nogués-Bravo1, D., … & Rockström, J. (2023). Earth beyond six of nine planetary boundaries. *Science Advances*, *9*, 1–16.

Silverstein, K. (2023, May 22). What is the human cost of global warming? *Forbes*. www.forbes.com/sites/kensilverstein/2023/05/22/what-is-the-human-cost-of-global-warming/

Steffen, W., Richardson, K., Rockström, J., Cornell, S.E., Fetzer, I., Bennett, E.M., Biggs, R., Carpenter, S.R., de Vries, W., de Wit, C.A., Folke, C., Gerten, D., Heinke, J., Mace, G.M., Persson, L.M., Ramanathan, V., Reyers,B., & Sverker Sörlin, S. (2015). Planetary boundaries: Guiding human development on a changing planet. *Science*, *347*(6223), 1259855-1–1259855-10.

Steffen, W., Rockström, J., Richardson, K., Lenton, T.M., Folke, C., Liverman, D., Summerhayes, C.P., Barnosky, A.D., Cornella, S.E., Crucifixi, M., Donges, J.F., Fetzera, I., Lade, S.J., Marten Scheffer, M., Winkelmann, R., & Schellnhuber, H.J. (2018). Trajectories of the earth system in the Anthropocene. *Proceedings of the National Academy of Sciences*, *115*(33), 8252–8259.

Steffen, W., & Morgan, (2021). From the Paris Agreement to the Anthropocene and planetary boundaries framework: An interview with Will Steffen. *Globalizations*, *18*(7), 1298–1310. https://doi.org/10.1080/14747731.2021.194007

Tschofen, P., Azevedo, I.L., & Muller, N.Z. (2019). Fine particulate matter damages and value added in the US economy. *Proceedings of the National Academy of Sciences*, *15*(40), 19857–19862. www.pnas.org/cgi/doi/10.1073/pnas.1905030116

United Nations. (2021, April 22). *Climate change link to displacement of most vulnerable is clear: UNHCR. UN News*. https://news.un.org/en/story/2021/04/1090432

United Nations. (2025). *SDG goals: Sustainable cities and communities*. United Nations Statistics Division Development Data and Outreach Branch. https://unstats.un.org/sdgs/report/2023/goal-11/#:~:text=Approximately%201.1%20billion%20people%20currently,convenient%20access%20to%20public%20transportation

United Nations Framework Convention on Climate Change. (2022, April 13). *What is the triple planetary crises?* https://unfccc.int/news/what-is-the-triple-planetary-crisis

United Nations Department of Economic and Social Affairs. (2020). *World social report 2020: Inequality in a rapidly changing world.* https://desapublications.un.org/publications/world-social-report-2020-inequality-rapidly-changing-world

United Nations. World Social Report 2024. (2024). www.un.org/development/desa/dpad/wp-content/uploads/sites/45/DESA-World-Social-Report_2024_FINAL_WEB.pdf

United Nations Environment Programme [UNEP]. (2022). *Spreading like wildfire—The rising threat of extraordinary landscape fires. A UNEP rapid response assessment.* www.unep.org/resources/report/spreading-wildfire-rising-threat-extraordinary-landscape-fires

United Nations High Commissioner for Refugees. (2021). *Mid-year trends 2021.* www.unhcr.org/sites/default/files/legacy-pdf/618ae4694.pdf

U.S. Global Leadership Coalition. (2021). *COVID-19 Brief: Impact on the Economies of Low-Income Countries.* www.usglc.org/coronavirus/economies-of-developing-countries/

van Eeden, L.M., Nimmo, D., Mahony, M., Herman, K., Ehmke, G., Driessen, J., O'Connor, J., Bino, G., Taylor, M. & Dickman C.R. (2020). *Impacts of the unprecedented 2019-2020 bushfires on Australian animals.* World Wildlife Fund for Nature—Australia. www.wwf.nl/globalassets/pdf/wwf-impacts-of-the-unprecedented-2019-2020-bushfires-on-australian-animals.pdf

Wilkinson, J.L., Boxall, A.B.A., Kolpin, D.W., Leung, K.M.Y., Lai, R.W.S., Galban-Malagon, C., Adell, A.D., Mondon, J., Metian, M., Marchant, R.A., Bouzas-Monroy, A., Cuni-Sanchez, A., Coors, A., Carriquiriborde, P., Rojo, M., Gordon, C., Cara, M., Moermond, M., Luarte, T., … & Teta, C. (2022). Pharmaceutical pollution of the world's rivers. *Proceedings of the National Academy of Sciences, 119*(8), 1–10.

Winship, S., Pulliam, C., Shiro, A.G., Reeves, R.V., & Deambrosi, S. (2020). *Long shadows: The black-white gap in multigenerational poverty.* The American Enterprise Institute and the Brookings Institute. www.brookings.edu/wp-content/uploads/2021/06/Long-Shadows_Final.pdf

World Bank Group. (2020). *Reversals of fortune. Poverty and shared prosperity 2020.* https://openknowledge.worldbank.org/server/api/core/bitstreams/611fc6f2-140b-551e-9371-468eec64c552/content

World Health Organization. (2024). *Number of COVID-19 deaths reported to WHO (cumulative total), world.* https://data.who.int/dashboards/covid19/deaths?n=c

World Meteorological Organization. (2022). *State of the global climate 2021.* https://library.wmo.int/viewer/56300/download?file=1290_Statement_2021_en.pdf&type=pdf&navigator=1

World Wildlife Fund [WWF]. (2024). Living planet report 2024—*A* system in peril. https://files.worldwildlife.org/wwfcmsprod/files/Publication/file/5gc2qerb1v_2024_living_planet_report_a_system_in_peril.pdf

The Massive Consumption of Natural Resources

3

"By living so far beyond our environmental means, and running up ecological debts, we make two mistakes. First, we deny millions globally who already lack access to sufficient land, food and clean water the chance to meet their needs. Secondly, we put the planet's life support mechanisms in peril".

Andrew Simms[1], Economist, New Weather Institute

OVERVIEW

Chapter 2 offered an overview of the anthropogenic destruction impacting people and the natural environment. This chapter continues with the hyperutilization of our natural resources globally while drawing attention to the resource utilization of the United States (U.S.). It will conclude with an overview of the varying theories that discern how we came to this precarious situation.

WHO IS DOING THE EMITTING?

As of January 15, 2025, the global population surpassed over 8 billion people while the population of the U.S. comprised over 341 million people (U.S.

DOI: 10.1201/9781003153528-4

Census Bureau, 2025). As noted in Figure 3.1, the U.S. is the world's second leading emitter of carbon dioxide (CO_2) emissions while China, with a population of over 1.4 billion people and a global manufacturing hub, is first (Global Carbon Atlas, 2025). This points to the hyper-consumption of resources in the U.S. India ranks third in global $C0_2$ emissions with a population of over 1.3 billion people (Global Carbon Atlas, 2025). The Russian Federation follows and Japan rounds out the top 5 global carbon emitters (Global Carbon Atlas, 2025).

The U.S. is a member of the group of the world's largest economies, the Group of Twenty (G20), which accounts for 77% of global greenhouse emissions (GHGE) (Australian Department of Foreign Affairs and Trade [ADFAT], 2024; United Nations Environment Programme [UNEP], 2024). In contrast, the least developed countries in the world account for 3% (UNEP, 2024). The G20 Countries include Argentina, Australia, Brazil, Canada, China, France, Germany, India, Indonesia, Italy, Japan, the Republic of Korea, Mexico, Russia, Saudi Arabia, South Africa, Turkey, United States of America, and the European Union with Spain as a permanent guest (ADFAT, 2024). Collectively, G20 emissions rose in 2023 and commitments to reduce GHGE fell short (UNEP, 2024). Without demonstrable actions, the U.S. is the "least likely" of the G20 countries to achieve its goals of lowering its Nationally Determined Contribution (NDC) (UNEP, 2024). To quote Inger Andersen, Executive Director of the United Nations Environment Programme (UNEP, 2024, p. XI):

> Essentially, we would need global mobilization on a scale and pace never seen before. Many will say this is impossible. But to focus solely on whether it is possible misses one crucial point: the transformation to net-zero economies must happen, and the sooner this global transformation begins the better. Every fraction of a degree avoided counts in terms of lives saved, economies protected, damages avoided, biodiversity conserved and the ability to rapidly bring down any temperature overshoot.

Seventy percent (70%) of cumulative historical global industrial carbon emissions between the years 1854 to 2022 can be traced to 122 industrial entities (InfluenceMap, 2024). Despite the increased use of alternative energies in recent years, two-thirds of global energy needs were fulfilled with fossil fuels in 2023 (International Energy Agency, 2024). Furthermore, where a child is born can notably impact their respective lifetime use of natural resources—as well as that of their offspring (Murtaugh & Schlax, 2009). To start, becoming a parent altogether increases one's respective carbon legacy and possible continued contribution to global carbon emissions (Murtaugh & Schlax, 2009). Children born in the U.S. accounted for the highest carbon

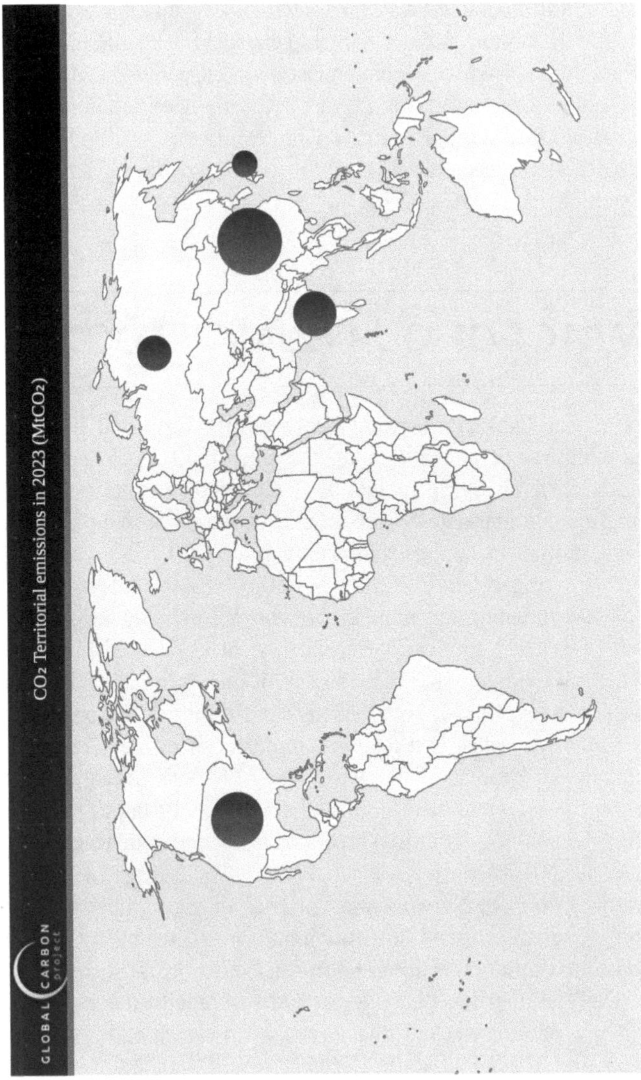

FIGURE 3.1 Top 5 Countries Carbon Dioxide Emissions ($MtCO_2$), Global Carbon Project, 2023.

legacies globally, far surpassing other countries such as Japan (the U.S. is 4.6 times greater), China (the U.S. is 7 times greater), Nigeria (the U.S. is 90 times greater), and Bangladesh (the U.S. is 168 times greater), respectively (Murtaugh & Schlax, 2009).

The top 10% of income earners across the globe generated 36% of carbon emissions while the bottom 50% of earners generated 15% that includes those that are extremely poor (4% of global emissions), respectively (Hubacek et al., 2017). Researchers find that if we are to swiftly reduce carbon emissions as a global society, a substantial reduction in resource consumption amongst top income earners is necessary along with the degrowth of rich economies (Hubacek et al., 2017; Higgs, 2020).

THE IMPACT OF OUR HUMAN FOOTPRINT

Every facet of our lives come from the resources furnished by the living systems and the living beings of the planet (Merhoff, 2019). When calculating both the biomass of all living matter (that included humans and livestock) and all human-made inanimate objects in use, researchers found that over the past 100 years, human-made mass has doubled every 20 years where in 2020, it surpassed our living biomass (Elhacham et al., 2020). Moreover, humans and livestock far outweigh the number of wild animals on the planet while domesticated birds used for food outweigh the mass of wild birds (Bar-On et al., 2018). As Wiedmann et al. (2015a) point out, "Humanity is using natural resources at a level never seen before"; our material footprint is unprecedented. The United States is the largest importer of primary resources and China is the largest exporter; China was assessed to have the largest material footprint while the U.S. had the second largest material footprint (Wiedmann et al., 2015a). Developing countries account for smaller material footprints (Wiedmann et al., 2015b).

The Global Footprint Network uses metrics to assess how much nature we have as a global society and how much nature is consumed by humans—including waste assimilation (Global Footprint Network, 2024a). They define "Overshoot Day" as the day of the year when humanity has exhausted the Earth's budget of naturally renewable resources in the calendar year (Global Footprint Network, 2024b). Over the past 50 years, Overshoot Day has generally occurred earlier every year since December 25, 1971; August 1st marked Overshoot Day in 2024, denoting unsettling biocapacity deficits (Global Footprint Network, 2024b). The Global Footprint Network offers a "footprint calculator" (footprintcalculator.org) where individuals can understand their respective ecological footprint and henceforth reduce their footprint (Global Footprint Network, 2024c).

DISCERNING ANTHROPOGENIC ENVIRONMENTAL DEGRADATION

The proposed geological epoch called "the Anthropocene" is widely recognized by professionals in planetary health, healthcare and medicine, public health, and the general public (Zalasiewicz et al., 2021). The Anthropocene targets the impact of fossil fuels in the degradation of and destabilization of the Earth System (Zalasiewicz et al., 2021). The International Commission on Stratigraphy rejected the concept of the Anthropocene in March 2024 but noted the importance of its use (International Commission on Stratigraphy, 2023). Other frameworks across a multitude of disciplines have also discerned the anthropogenic degradation of the Earth System such as the Chthulacene, Plantationocene, Capitalocene, Polemocene, and Urbanocene; there are at least 80 frameworks and several will be summarized hereafter (Chwałczyk, 2020).

Supporters of the Capitalocene framework find common ground with the Anthropocene but argue that it ignores capitalistic systems and the massive accumulation of wealth and power acquired by way of direct violence, class exploitation, and alienation (Moore, 2017). The Capitalocene degrades both humans and nature as "cheap objects" (Moore, 2017). Other scholars recognize the impact of war in environmental degradation where warfare and capitalism are intertwined, proposing the Polemocene (Antonucci, 2021). Antonucci (2021) explains that, "…capital accumulation has commenced so spectacularly because of destructive wars", resulting in "…endless accumulation". The Plantationocene describes the impact of racism in the assault on the natural environment (Murphy & Schroering, 2020). The Urbanocene underscores the role of the growth and intensification of urban areas in the ecological crisis (Chwałczyk, 2020). Finally, the Chthulucene emphasizes a future of repairing ecological damage and striving for multispecies well-being and ecojustice for all inhabitants on Earth (Haraway, 2015). More specifically, Haraway (2016) writes, "The order is reknitted: human beings are with and of the Earth, and the biotic and abiotic powers of this Earth are the main story".

While the reader can discern the merits of the proposed paradigms, all of the frameworks present the complexities imposed upon biodiversity, humans, and the Earth System. Importantly, we will not achieve the Sustainable Development Goals or reverse environmental degradation without restructuring systems and policies to create equity for the entirety of life on the planet (Wiedmann, 2015a; Higgs, 2020).

NOTE

1 BBC News. (2006, October 9). *Planet enters 'ecological debt'*. http://news.bbc.co.uk/2/hi/science/nature/6033407.stm

REFERENCES

Antonucci, J.P. (2021). Periodizing the Capitalocene as Polemocene militarized ecologies of accumulation in the long sixteenth century. *Journal of World-Systems Research*, *27*(2), 439–467. https://doi.org/10.5195/JWSR.2021.1045

Australian Department of Foreign Affairs and Trade [ADFAT]. (2024). *The G20*. www.dfat.gov.au/trade/organisations/g20

Bar-On, Y.M., Phillips, R., & Milo, R. (2018). The biomass distribution on Earth. *Proceedings of the National Academy of Sciences*, *115*(25), 6506–6511. www.pnas.org/cgi/doi/10.1073/pnas.1711842115

Chwałczyk, F. (2020). Around the Anthropocene in eighty names—Considering the Urbanocene proposition. *Sustainability*, *12*(4458), 1–33. https://doi.org/10.3390/su12114458

Elhacham, E., Ben-Uri, L., Grozovski, J., Bar-On, Y.M., & Milo, R. (2020). Global human made mass exceeds all living biomass. *Nature*, *588*, 1–16. https://doi.org/10.1038/s41586-020-3010-5

Global Carbon Atlas. (2025, January 15). *World total: 37792 MtCO₂*. https://globalcarbonatlas.org/emissions/carbon-emissions/

Global Footprint Network. (2024a). *Ecological footprint*. www.footprintnetwork.org/our-work/ecological-footprint/

Global Footprint Network. (2024b). *Earth overshoot day*. www.footprintnetwork.org/our-work/earth-overshoot-day/#:~:text= In%202024%2C%20Earth%20Overshoot%20Day,carbon%20dioxide%20in%20the%20atmosphere.

Global Footprint Network. (2024c). *What is your ecological footprint?* www.footprintcalculator.org/home/en

Haraway, D. (2015). Anthropocene, capitalocene, pantationocene, chthulucene: Making Kin. *Environmental Humanities*, *6*, 159–165.

Haraway, D. (2016). Tentacular thinking: Anthropocene, Capitalocene, Chthulucene. *e-flux Journal*, *75*, 1–13.

Higgs, K. (2020). How sustainable are the SDGs? In Washington, H. (Ed.). *Ecological economics: Solutions for the future* (pp. 109–130). Self-published.

Hubacek, K., Baiocchi, G., Feng, K., & Patwardhan, A. (2017). Poverty eradication in a carbon constrained world. *Nature Communications*, *8*(912), 1–9. https://doi.org/10.1038/s41467-017-00919-4

InfluenceMap. (2024, April). *The carbon majors database: Launch report*. https://carbonmajors.org/briefing/The-Carbon-Majors-Database-26913

International Commission on Stratigraphy. (2023). *Joint statement by the IUGS and ICS on the vote by the ICS subcommission on quaternary stratigraphy*. https://stratigraphy.org/news/152

International Energy Agency. (2024). *World energy outlook*. https://iea.blob.core.windows.net/assets/60db7f34-da7f-41d7-8232-c0b718463fd0/WorldEnergyOutlook2024.pdf

Merhoff, L.A. (2019). Biological diversity—Life on Earth. In Hawkins, I.W. (Ed.) (2019). *Promoting biodiversity in food systems* (pp. 13–28). CRC Press. https://doi.org/10.1201/b22084

Moore, J.W. (2017). The Capitalocene, part I: On the nature and origins of our ecological crisis. *The Journal of Peasant Studies*, *44*(3), 594–630. http://dx.doi.org/10.1080/03066150.2016.1235036

Murphy, M.W., & Schroering, C. (2020). Refiguring the Plantationocene: Racial capitalism, world-systems analysis, and global socioecological transformation. *Journal of World-Systems Research*, *26*(2), 400–415. https://doi.org/10.5195/jwsr.2020.983

Murtaugh, P.A. & Schlax, M.G. (2009). Reproduction and the carbon legacies of individuals. *Global Environmental Change, 19*, 14–20.

United Nations Environmental Programme. (2024). *Emissions Gap Report 2024: No more hot air... please! With a massive gap between rhetoric and reality, countries draft new climate commitments*. www.unep.org/resources/emissions-gap-report-2024

U.S. Census. (2025, January 15). *U.S. and World population clock*. www.census.gov/popclock/

Wiedmann, T.O., Schandl, H., Lenzen, M., Moran, D., Suh, S., West, J., & Kanemotoc, K. (2015a). The material footprint of nations. *Proceedings of the National Academy of Sciences*, *112*(20), 6271–6276. www.pnas.org/cgi/doi/10.1073/pnas.1220362110

Wiedmann, T.O., Schandl, H., Lenzen, M., Moran, D., Suh, S., West, J., & Kanemoto, K. (2015b). Supporting information (supplement). Proceedings of the National Academy of Sciences, 1–9. https://doi.org/10.1073/pnas.1220362110

Zalasiewicz, J., Waters, C.N., Ellis, E.C., Head, M.J., Vidas, D., Steffen, W., Thomas, A.J., Horn, E., Summerhayes, C.P., Leinfelder, R., McNeill, J.R., Gałuszka, A., Williams, M., Barnosky, A.D., Richter, D.B, Gibbard, P.L., Syvitski, J. Jeandel, C., Cearreta, A., … & Zinke, J. (2021). The anthropocene: Comparing its meaning in geology (Chronostratigraphy) with conceptual approaches arising in other disciplines. *Earth's Future*, *9*(3), 1–25. https://doi.org/10.1029/2020EF001896

4 Global Industrialized Food Systems and Environmental Degradation

"The current food system is neither healthy, nor sustainable."
Marco Springmann[1], PhD & Florian Freund[1], PhD,
Planetary Health Researchers

OVERVIEW

Today, the globalization of the food supply including large-scale industrial farms, intensive monocropping, and confined animal and aquaculture operations are commonplace. International policies and agrifood commerce alongside national policies have bypassed traditional ecological knowledge, embracing rapid and lucrative food production. We can now assess the impact of food and food systems on the natural environment as well as the cost to human health through scientific data and assessments of chronic disease morbidity and mortality. This chapter will address the grave public health concerns that connect food systems, industrial agriculture, and animal products to our shared destiny with the Earth System. This includes how food systems and food choices contribute to breaching our planetary boundaries including plastics food packaging; food insecurity and nutrient losses—as well as the impact of agrichemicals, zoonotic disease, and antibiotic resistance.

DOI: 10.1201/9781003153528-5

CLIMATE EXTREMES, BREACHING OUR PLANETARY BOUNDARIES, AND FOOD INSECURITY

As mentioned in Chapter 2, the food system is connected to all six planetary boundaries that have been transgressed where scientists warn that we are headed towards the "hothouse Earth trajectory" where if not reversed, impaired biogeophysical feedbacks along with potential tipping points suggest irreversible and catastrophic changes (Steffen et al., 2018). Globally, greenhouse gas emissions (GHGE) and record setting heat events have reached new crescendos, exacerbating climate change and climate disasters with the year 2024 being the warmest year on record (World Meteorological Association, 2025).

Food systems are responsible for 34% of all GHGE with 27% coming from industrialized countries where six economies (China, Indonesia, U.S., Brazil, European Union, and India) were responsible for 51% of emissions (Crippa et al., 2021). Livestock accounted for 35% of methane emissions globally where methane is a greenhouse gas 30 times more potent than carbon dioxide (Crippa et al., 2021). Rice is also implicated in methane emissions and is an important staple food in Asian cultures (Crippa et al., 2021). Reducing global methane emissions would not only help mitigate the impact of GHGE—it would improve air quality and reduce premature all-cause mortality including respiratory and cardiovascular deaths (West et al., 2006).

Nitrous oxide, a greenhouse gas 300 times more potent than carbon dioxide, remains in the atmosphere for long periods and accounted for 9–14% of food system emissions (Crippa et al., 2021). Because nitrous oxide emissions have increased by 30% in recent decades and is implicated in the destruction of the ozone layer—anthropogenic sources are concerning where agriculture is responsible for the majority of emissions overall (Tian et al., 2020). Fertilizers are a major source of anthropogenic nitrous oxide emissions along with livestock manure (Tian et al., 2020). In Brazil alone, livestock manure used for fertilizer increased nitrous oxide emissions by 120% between 1980–2016 (Tian et al., 2020). Aquaculture has also been recognized as an important source of nitrous oxide emissions, too (Hu et al., 2012).

The immense amount of waste from the billions of food-animals slaughtered each year cannot be readily assimilated into the natural environment which poses an array of hazards to both humans and ecosystems such as but not limited to air pollution, water pollution, and antibiotic resistance (American Public Health Association, 2019). In fact, the American Public Health Association in the U.S. has called for a moratorium on concentrated animal feeding operations that will be discussed in Chapter Eight (American Public Health Association, 2019).

CLIMATE EXTREMES AND THE AVAILABILITY OF THE FOOD SUPPLY

Warmer waterways not only increase the severity of extreme weather events as seen with hurricanes and tropical storms—but are linked to toxic algal blooms and parasites that can kill or impair fish and other marine animals. In 2019, thousands of salmon died in Alaska in the Yukon River before they were able to spawn due to an unprecedented heat wave that warmed rivers (Perrone, 2019). Cattle are also sensitive to heat distress which impacts meat and milk production, fertility, growth, and product yields (Thorton et al., 2022). A modeling study of heat stress on global cattle production systems under low-and-high GHGE scenarios found cattle production losses at nearly $40 billion to $150 billion annually, respectively (Thorton et al., 2022). Additionally, livestock may not be able to survive in low-income countries where regions may become altogether too hot and humid (Thorton et al., 2022).

A global systematic review determining how environmental changes such as limited water availability, increased carbon dioxide levels, and water salinity would impact the production of fruits, nuts, and seeds found that both yields and nutritional quality would be negatively impacted, increasing the risk of food insecurity and limiting nutrient availability (Alae-Crew et al., 2020). Additionally, the protein, vitamin, and micronutrient content of rice will decrease as a result of increased atmospheric carbon dioxide levels which could negatively impact over 600 million people in low and middle-income Asian countries (Zhu et al., 2018). Hence, the data demonstrates the need for expediency in mitigating climate change and ending environmentally destructive agricultural practices so to maintain food and nutrient security.

ZOONOTIC DISEASE AND ANTIBIOTIC RESISTANCE

Zoonotic Disease

The COVID-19 pandemic illuminated the connections between food systems, zoonosis, and environmental degradation. Importantly, scientists identify the emergence of the novel coronavirus (COVID-19) as, "…a widely predicted consequence of how people source food, trade animals, and alter environments"

(United Nations Environment Programme [UNEP] & International Livestock Research Institute [ILRI], 2020). As we share the Earth system with non-human living beings, we share pathogens with them, too. As such, pathogens can be transmitted to humans by animals which is called zoonotic spillover (Ellwanger & Chies, 2021). Most infectious diseases that humans experience including emerging infectious diseases are zoonotic, meaning they are spread by animals (UNEP & ILRI, 2020; Ellwanger & Chies, 2021).

Assessing the role of the natural environment in the COVID-19 pandemic found several underlying drivers of zoonotic disease: the increased demand for animal protein and subsequent intensive livestock/factory farming; the increased exploitation of wild animals; climate change; the far-reaching impact of transportation; land-use changes and extractive practices that denude the natural environment; and a lack of policies and regulations to mitigate zoonotic spillover (UNEP & ILRI, 2020).

Healthy and biodiverse ecosystems have an abundance of competing host species which decreases the risk of disease transmission to humans while also shielding interactions between humans and wildlife (Ellwanger, 2021). While converting forests for agricultural use is seemingly beneficial, altering such ecosystems and the associated biodiversity increases the risk of emerging zoonotic disease and zoonotic hazards (pathogens and parasites) compared to areas that were not disturbed (Ellwanger & Chies, 2021; Gibb et al., 2020).

Industrial food-animal production is also linked to the highly pathogenic avian influenza which has killed more than 500 million farmed birds globally since 2021 (Kuiken & Cromie, 2022; Compassion in World Farming, 2023). The pathogen has also spilled over into wild bird populations where millions of birds have been impacted (Klaassen & Wille, 2023). Researchers Klaassen and Wille (2023) noted that, "This panzootic did not emerge from nowhere, but rather is the result of 20 years of viral evolution in the ever-expanding global poultry population". Also alarming are reports of highly pathogenic avian flu spillover to gray and harbor seals in the New England region of the U.S. (Puryear et al., 2023).

Antibiotic Resistance

The use and misuse of antibiotics in animal agriculture drives antibiotic resistance, a top global public health threat (World Health Organization [WHO], 2023). The widespread administration of subtherapeutic levels of medically important antibiotics/antimicrobials in food-animal production and aquaculture enables pathogens to survive and evolve, rendering some antimicrobials less effective or useless; this creates unnecessary mortality around the globe (Antimicrobial Resistance Collaborators, 2022; Tang et al., 2017; WHO, 2023).

Using predictive statistical modeling, over 4.95 million deaths globally were attributed to antimicrobial resistance, disproportionately impacting low-income countries (Antimicrobial Resistance Collaborators, 2022). Global use of antimicrobials is projected to increase by 67% during the years 2010–2030 while there's been an alarming increase in common infections that are resistant to antimicrobials (Van Boeckel et al., 2015; WHO, 2021).

The World Health Organization (WHO) has recommended discontinuing the prophylactic use of antibiotics for food-animals; the data demonstrates that when antibiotics are withheld, antibiotic resistance decreases (WHO, 2017). Interestingly, Shepon et al. (2023) used a scenario framework to understand the relationship between agricultural practices, dietary patterns, food system-related drivers of zoonotic spillover, and the relative risk of zoonotic spill-over. Their analysis found that a plant-based dietary pattern grown using agroecological practices along with low amounts of animal product consumption would improve planetary health outcomes and lower the risk of zoonotic spillover, antibiotic resistance, and biodiversity loss compared to industrialized farming systems that use agrichemicals and antibiotics for producing animal-based dietary patterns and processed foods (Shepon et al., 2023).

In the U.S., antibiotics used in human medicine has decreased in recent years while increasing in industrial food-animal production (Wallinga, 2021). U.S. federal policies have failed to curb the use of antibiotics in food-animal production, a chief driver of antibiotic resistance (Wallinga et al., 2022). In 2017, the treatment of community-onset and hospital-acquired infections pathogens amongst patients in the U.S. Veterans Affairs healthcare system cost an estimated $4.6 billion dollars per year (Nelson et al., 2021). The Centers for Disease Control and Prevention (CDC) reports that 48,000 people die and over 3 million experience antimicrobial infections each year in the U.S. (Centers for Disease Control and Prevention [CDC], 2019). Thus, the data speaks to expediently changing international and federal policies to restrict the widespread use of antibiotics in food-animal production so to save lives, money—and the precious life-saving medicines themselves.

Antimicrobials are also used as agrichemicals in plant agriculture (Miller et al., 2022). Antibiotic-resistant genes impact the food chain by persisting in soil, air, waterways, crops, animal waste, sludge, sewage—and the animal products themselves (Skandalis et al., 2021). A case-controlled study of women in the U.S. with urinary tract infections found the multi-drug-resistant pathogen *Escherichia coli* (*E. coli*) amongst those that frequently consumed chicken and those that moderately consumed pork (Manges et al., 2007). Analyzing urinary cultures of both men and women with *E. coli* urinary tract infections in the Netherlands, higher intakes of pork and chicken were associated with drug-resistance (Mulder et al., 2019). Moreover, researchers in Flagstaff, Arizona used genetic analysis techniques to identify that 8% of meats purchased at local grocery stores were linked to *E. coli* urinary tract infections amongst patients at the Flagstaff Medical Center during the year 2012 (Liu et al., 2023).

Aquaculture

Zoonotic disease from aquaculture is also concerning where the aquatic food-animal supply chain is overlooked in the transmission of antimicrobial-resistant bacteria and genes (Watts et al., 2017). Similar to terrestrial factory farming, aquaculture facilities may house large quantities of fish in small bodies of water where the risk of antimicrobial resistance and disease transmission increases and can remain in fish tissues, too (Watts et al., 2017). There is no unified surveillance system for the aquatic species that humans consume.

A systematic review examined multi-drug resistance and bacterial pathogens in aquaculture in 40 countries and found that most countries demonstrated high levels of antimicrobial resistance; there was also a relationship between climate change, warmer waters, and antibiotic resistance (Reverter et al., 2020). Additionally, there was higher mortality amongst fish infected with pathogens, too (Reverter et al., 2020). An analysis spanning nearly two decades in Asia found elevated rates of antibiotic-resistant bacteria in aquatic food-animals (Schar et al., 2021). Ultimately, scientists strongly recommend the swift implementation of global surveillance systems as well as strategies that preserve the therapeutic value of antimicrobials by minimizing or eliminating their application in aquaculture (Reverter et al., 2020; Schar et al., 2021; Watts et al., 2017).

AGRICHEMICALS

Agrichemicals are widely used in industrialized agriculture where they disrupt the biological processes of not only the target organism—but can embody negative externalities impacting ecosystems, biodiversity, and human health (Tudi et al., 2021; Gunstone et al., 2021; Hofmann et al., 2021). In the U.S., agrichemicals are a leading source of waterway and groundwater contamination (U.S. Geological Survey [USGS], 2019). The mass use of nitrogen and phosphorus in industrial agriculture as nutrients for crops creates water pollution and can destroy ecosystems while creating dead zones—and has transgressed our planetary boundary for biogeochemical flows as delineated in Chapter Two (Frei et al., 2021; Richardson et al., 2023; Donner, 2006). Some of the largest consumers and producers of agrichemicals globally are the European Union, Brazil, U.S., and China where together, they utilized 6.8 billion pounds of pesticides in 2016 (Donley, 2019). In 2016, 388 million pounds of pesticides that were banned in the European Union were used in the U.S. (322 million pounds), China (40 million pounds), and Brazil (26 million pounds), respectively (Donley, 2019).

Genetically modified crops utilize more agrichemicals (Benbrook, 2012). Glyphosate, the most widely utilized agrichemical in the world can accumulate in waterways and has drawn considerable attention including being considered "probably carcinogenic" by the International Agency for Research on Cancer (International Agency for Research on Cancer [IARC], 2015). A wide array of research on glyphosate use has found: accumulation in waterways; changes in human cellular functions; and possible indirect increases in pathogenic bacteria (Davoren & Shiestel, 2018). Benbrook (2019)delineated the differences in how the IARC and the U.S. Environmental Protection Agency (EPA) determined the safety of glyphosate, concluding, "IARC's evaluation relied heavily on studies capable of shedding light on the distribution of real-world exposures and genotoxicity risk in exposed human populations, while EPA's evaluation placed little or no weight on such evidence" (Benbrook, 2019).

Unintentional Acute Pesticide Poisoning is another major public health issue (Boedecker et al., 2020). A systematic review of the global burden of Unintentional Acute Pesticide Poisoning (UAPP) during the years 2006–2018 in 141 countries found over 7,400 deaths and over 733,000 non-fatal cases (Boedecker, 2020). From there, researchers estimated that 385 million non-fatal cases and 11,000 fatal cases occur annually around the globe (Boedecker, 2020). Furthermore, Hoffman et al. (2021) found an association between the use of the insecticide permethrin and multiple myeloma amongst farmers in Iowa and North Carolina in the prospective Agricultural Health Study. Bao et al. (2020) used data from the U.S. National Health and Nutrition Examination Study (NHANES) to determine that pyrethroids (insecticides) were associated with an increased risk of all-cause mortality and cardiovascular disease mortality amongst a representative sample of the general population aged 20–60. Additionally, a meta-analysis found that exposure to glyphosate-based herbicides was linked to Non-Hodgkin Lymphoma (Zhang et al., 2019).

FOOD PACKAGING AND PLASTICS POLLUTION

While Chapter Two introduced the impact of plastics on ecosystems, biodiversity, and the planetary boundary novel entities, this section will draw attention to connections between plastics pollution, agriculture, the food system, and human health. A large body of research points to the deleterious impact of plastics and food packaging on the natural environment as well as concern for the human body as plastic is made from ~13,000 industrial chemicals and toxicants (Landrigan et al., 2023; Guyer et al. 2017; United Nations Environment Programme [UNEP] & Secretariat of the Basel, Rotterdam and Stockholm Conventions, 2023). While the 2022 United Nations Environment Programme (UNEP) Resolution "End Plastic Pollution: Towards an International Legally Binding Instrument" may offer a pathway to stem global plastics pollution, current food system and agricultural practices are deeply entrenched in plastic use (United Nations Environment Assembly of the UNEP, 2022). A comprehensive global assessment of the annual impact of plastic on planetary health, human health, and the economy found that in 2015 alone, health-related costs exceeded $250 billion dollars (Landrigan et al., 2023).

The Scope of Plastics Pollution

The burden of plastics pollution including microplastics pollution is extraordinarily widespread. From 1950 to 2015, an estimated 8,300 million metric tons of virgin plastics have been produced globally and by 2050, an estimated 12,000 million metric tons will enter landfills or the natural environment (Geyer et al., 2017). Plastics manufacturing is a major industry in the U.S. where in 2020, the U.S. imported nearly $59 billion in plastics products and exported $60.2 billion (National Academies of Sciences, Engineering, and Medicine [NAS], 2022). In fact, utilizing data from 217 countries, researchers calculated that in the year 2016, the U.S. topped the world in plastic waste generation (Law, 2020). Recycling plastic is a global challenge where only 9% is recycled and only 10% has been recycled more than once (Geyer et al., 2017). Less than 9% of plastics were recycled in the U.S. in 2018 (U.S. EPA, 2020). Ocean plastics pollution is deeply concerning; by 2050, it's been estimated that there will be more plastic in the ocean by weight than fish (World Economic Forum et al., 2016).

Plastics are generally comprised of fossil fuel-based materials along with a heterogeneous and proprietary blend of chemicals; many are considered hazardous (NAS, 2022; UNEP & Secretariat of the Basel, Rotterdam and Stockholm Conventions, 2023; Rochman et al., 2019).

Plastics constitute a widerange of products including stabilizers, additives, colorants, flame retardants, sealants, etc. (Health And Environment Alliance [HAEA], 2020). Producing plastics contributes approximately 4% of the overall burden of global GHGE, where the financial burden of GHGE attributed to plastics was calculated at $341 billion annually (Landrigan et al.,

2023). Additionally, when plastics are exposed to sunlight and water, they break down and become another important yet overlooked source of GHGE that may be released throughout a plastic product's entire lifecycle (HAEA, 2020; Royer, 2018).

The food system is heavily implicated in plastics pollution. A global classification of macro-litter categories in waterways (with the exception of the open ocean where debris was associated with ocean-based activities) found that 80% was made of plastic single-use food service implements (Morales-Caselles et al., 2021). The healthcare industry is an important contributor to plastics pollution where it's estimated that the U.S. generates 1.7 million tons per year annually (Rizan et al., 2020). Rizan et al. (2020) states, "The frequent and growing use of disposable plastic products due to potential but largely unsubstantiated infection risk, or convenience, cannot be used by healthcare to exempt itself from the broad global imperative to reduce reliance on plastics".

An example of how plastics pollution plays out daily around the globe comes from Hawaii's Kamilo Beach which struggles with astonishing amounts of debris brought in from the ocean currents where 96% of the 15-20 tons of the ocean pollution that accumulates there annually is plastic (Barney & Van Dyke, 2020). Microplastics (fragments of plastic so small they cannot be seen with the human eye) are now found in the deepest points of the ocean where it has accumulated in the bodies of amphipods living in the deep-sea (Jamieson et al., 2019). Furthermore, microplastics are recognized as a vector for pollutants, toxicants, pathogens, and antibiotic-resistant genes as well as horizontal gene transfer (Tuvo et al., 2023).

Plastics in Agriculture

The role of plastics in land-based agriculture ranges from biosolids, polymer coated fertilizers, seed coatings, irrigation, mulch film, silage films, protective nets, sheaths, pots and plug trays, greenhouse applications/uses, food storage, etc. (United Nations Environment Programme, 2022; Food and Agriculture Organization [FAO], 2021). Globally, over 12 million tons of plastics products were used in crop and food-animal production in 2019 along with over 37 million tons for food packaging (FAO, 2021). Livestock and crop production accounted for the greatest use at 10 million tons per year followed by fisheries (over 2 million tons) and aquaculture (0.2 million tons), respectively (FAO, 2021). Agricultural plastics are generally single-use items (FAO, 2021). They can contaminate soil, decrease soil productivity and agricultural yields while negatively impacting trophic levels (FAO, 2021). Agrichemicals and microplastics are considered a "toxic combination" as these synthetic chemicals not only persist in the natural environment destroying even the smallest elements and processes of life—but haven't been recognized for the entirety of hazards they can impart (Center for International Environmental Law, 2022).

Plastics and Chemicals in Food and Water Consumption

Food packaging is not only deleterious to the natural environment—but is also concerning for human health (Landrigan et al., 2023). Bi et al. (2023) analyzed plastic food packaging for organophosphate esters (OPE) which are chemicals used in applications such as plasticizers, flame retardants, etc. Using ultra high-performance mass spectrometry, these researchers found that OPE's had migrated from food packaging to the food product where they could be ingested (Bi et al., 2023). Cox et al. (2019) found that when evaluating a portion of Americans' caloric intake for microplastics, annual consumption ranged from 36,000 to 52,000 particles. Hussain et al. (2023) found that a significant migration of microplastics and nanoparticles were released when baby food containers and food pouches were heated in the microwave for three minutes; when conducting in vitro testing on human embryonic kidney cells, they found evidence of potential toxicity (Hussain et al., 2023).

Zimmerman et al. (2019) tested plastics commonly used as food receptacles and household goods including "bioplastics" for in vitro toxicity and found that most were implicated. High toxicity was found amongst some food wrappers and containers where some demonstrated antiandrogenic (testosterone blocking) and estrogenic properties (Zimmerman et al., 2019). Further testing of bioplastics found that they shouldn't be considered a safer alternative (Zimmerman et al., 2020).

Compostable paper food containers may be considered a solution to plastic take-out containers; however, some contained per-and polyfluoroalkyl substances (PFAS) that are forever chemicals (Panieri et al., 2022). The IARC considers the forever chemicals perfluorooctanoic acid (PFOA) and perfluorooctanesulfonic acid (PFOS) as carcinogenic and possibly carcinogenic to humans, respectively (Zahm et al., 2023).

Analyzing fast-food takeout receptacles in Canada using gamma ray emission spectroscopy, researchers found the highest levels of per- and polyfluoroalkyl substances (PFAS) in compostable paper bowls when compared to sandwich/burger wrappers, popcorn serving bags, and dessert and bread wrappers, respectively (Schwartz-Narbonne et al., 2023). An analysis of hundreds of plastic food packages from major retailers detected estrogenic activity with common usage and exposure to sunlight, boiling water—or when microwaved (Yang et al., 2011). Even packaging that was labeled asserting the positive connotation "bisphenol A (BPA) free" demonstrated estrogenic activity (Yang et al., 2011).

Often overlooked are the pipes that deliver water to homes and businesses. Older metal pipes are often replaced with plastic pipes including polyvinyl chloride (PVC) (Beyond Plastics, 2023). Because plastic pipes leach chemicals that may have consequences to human health such as endocrine disruption and the possible risk of cancer, copper pipes or recycled copper pipes are recommended (Beyond Plastics, 2023). Additionally, copper piping does not appear to substantially increase project costs (Beyond Plastics, 2023).

In closing, the industrialized food systems impact to the natural environment is staggering. Additionally, a large body of data demonstrates that chemicals and plastics are deeply entrenched in many facets of daily life which is also concerning for planetary health.

NOTE

1 Springmann, M., & Freund, F. (2022). Options for reforming agricultural subsidies from health, climate, and economic perspectives. *Nature Communications, 13*(82), 1–7. https://doi.org/10.1038/s41467-021-27645-2

REFERENCES

Alae-Carew, C., Nicoleau, S., Bird, F.A., Hawkins, P., Tuomisto, H.L., Haines, A., Dangour, A.D., & Scheelbeek, P.F.D. (2020). The impact of environmental changes onthe yield and nutritional quality of fruits, nuts and seeds: A systematic review. *Environmental Research Letters*, *15*, 1–13.

American Public Health Association. (2019, November 5). *Precautionary moratorium on new and expanding concentrated animal feeding operations*. American Public Health Association Policy Number: 20194. https://apha.org/policies-and-advocacy/public-health-policy-statements…oratorium-on-new-and-expanding-concentrated-animal-feeding-operations

Antimicrobial Resistance Collaborators. (2022). Global burden of bacterial antimicrobial resistance in 2019: A systematic analysis. *Lancet*, *399*, 629–655. https://doi.org/10.1016/S0140-6736(21)02724-0

Bao, W., Liu, B., Simonsen, D.W., & Lehmler, H.J. (2020). Association between exposure to pyrethroid insecticides and risk of all-cause and cause-specific mortality in the general US adult population. *JAMA Internal Medicine*, *180*(3), 367–374. https://doi.org/10.1001/jamainternmed.2019.6019

Barney, M., & Van Dyke, M.B. (2020, January 10). Welcome to Hawaii's 'plastic beach', one of the world's dirtiest places. *The Guardian*. www.theguardian.com/us-news/2020/jan/10/kamilo-beach-plastic-hawaii-pollution

Benbrook, C.M. (2019). How did the US EPA and IARC reach diametrically opposed conclusions on the genotoxicity of glyphosate-based herbicides? *Environmental Sciences Europe*, *31*(2), 1–16. https://doi.org/10.1186/s12302-018-0184-7

Benbrook, C.M. (2012). Impacts of genetically engineered crops on pesticide use in the U.S.– the first sixteen years. *Environmental Sciences Europe*, *24*(24), 1–13.

Beyond Plastics. (2023). *The perils of PVC plastic pipes.* Bennington College. https://static1.squarespace.com/static/5eda91260bbb7e7a4bf528d8/t/6491ce414930f2385aedb80c/1687277125680/The_Perils_of_PVC_Plastic_Pipes-April_2023_Digital.pdf

Bi, R., Meng, W., & Su, G. (2023). Organophosphate esters (OPEs) in plastic food packaging: Non-target recognition, and migration behavior assessment. *Environment International*, *177*, 1–10. https://doi.org/10.1016/j.envint.2023.108010

Boedecker W., Watts, M., Clausing, P., & Marquez, E. (2020). The global distribution of acute unintentional pesticide poisoning: Estimations based on a systematic review. *BMC Public Health*, *20*(1875), 1–19. https://doi.org/10.1186/s12889-020-09939-0

Center for International Environmental Law. (2022). *Sowing a plastic planet: How microplastics in agrochemicals are affecting our soils, our food, and our future.* CIEL. www.ciel.org/reports/microplastics-in-agrochemicals/

Centers for Disease Control and Prevention [CDC]. (2019). *CDC. Antibiotic Resistance Threats in the United States, 2019.* Atlanta, GA: U.S. Department of Health and Human Services, CDC.

Compassion in World Farming. (2023). *Bird flu: Only major farm reforms can end it.* www.ciwf.org.uk/research/animal-welfare/bird-flu-only-major-farm-reforms-can-end-it/

Cox, K.D., Covernton, G.A., Davies, H.L., Dower, J.F., Juanes, F., & Dudas, S.E. (2019). Human consumption of microplastics. *Environmental Science & Technology*, *53*, 7068–7074. https://doi.org/10.1021/acs.est.9b01517

Crippa, M., Solazzo, E., Guizzardi, D., Monforti-Ferrario, F., Tubiello, F.N., & Leip, A. (2021). Food systems are responsible for a third of global anthropogenic GHG emissions. *Nature Food*, *2*(3), 198–209. https://doi.org/10.1038/s43016-021-00225-9

Davoren, M.J., & Shiestel, R.H. (2018). Glyphosate-based herbicides and cancer risk: A post-IARC decision review of potential mechanisms, policy and avenues of research. *Carcinogenesis*, *39*(10), 1207–1215. https://doi.org/10.1093/carcin/bgy105

Donley, N. (2019). The USA lags behind other agricultural nations in banning harmful pesticides. *Environmental Health*, *18*(44), 1–12. https://doi.org/10.1186/s12940-019-0488-0

Donner, S.D. (2006). Surf or turf: A shift from feed to food cultivation could reduce nutrient flux to the Gulf of Mexico. *Global Environmental Change*, *17*, 105–113.

Ellwanger, J.H., & Chies, J.A.B. (2021). Zoonotic spillover: Understanding basic aspects for better prevention. *Genetics and Molecular Biology*, *44*(1 Suppl 1), 1–15.

Food and Agriculture Organization. (2021). *Assessment of agricultural plastics and their sustainability. A call for action.* United Nations. https://doi.org/10.4060/cb7856en

Frei R.J., Lawson, G.M., Norris, A.J., Cano, G., Vargas, M.C., Kujanpaa, E., Hopkins, A., Brown, B., Sabo, R., Barahney, J., & Abbott, B.W. (2021). Limited progress in nutrient pollution in the U.S. caused by spatially persistent nutrient sources. *PLoS ONE, 16*(11), e0258952. https://doi.org/10.1371/journal.pone.0258952

Gibb, R., Redding, D.W., Chin, K.Q., Donnelly, C.A., Blackburn, T.M., Newbold, T., & Jones, K.E. (2020). Zoonotic host diversity increases in human-dominated ecosystems. *Nature, 584*, 398–402. https://doi.org/10.1038/s41586-020-2562-8

Gunstone T., Cornelisse, T., Kendra Klein, K., Dubey, A., & Donley, N. (2021). Pesticides and soil invertebrates: A hazard assessment. *Frontiers in Environmental Science, 9*(643847), 1–21. https://doi.org/10.3389/fenvs.2021.643847

Guyer, R., Jambeck, J.R., & Law, K.L. (2017). Production, use, and fate of all plastics ever made. *Science Advances, 3*(e1700782), 1–5.

Health And Environment Alliance [HAEA]. (2020). *Turning the plastic tide: The chemicals in plastic that put our health at risk.* HAEA. www.env-health.org/turning-the-plastic-tide-the-chemicals-in-plastic-that-put-our-health-at-risk/#:~:text=The%20chemicals%20in%20plastic%20put%20our%20health%20at%20risk&text=Almost%20all%20plastics%20derive%20from,directly%20associated%20with%20plastics%20production.

Hoffman, J.N., Freeman, L.E.B., Murata, K., Andreotti, G., Shearer, J.J., Thoren, K., Ramanathan, L., Parks, C.G., Koutros, S., Lerro, C.C., Liu, D., Rothman, N., Lynch, C.F., Graubard, B.I., Sandler, D.P., Michael C. Alavanja, M.C., & Landgren, O. (2021). Lifetime pesticide use and monoclonal gammopathy of undetermined significance in a prospective cohort of male farmers. *Environmental Health Perspectives, 129*(1), 017003-1–017003-10. https://doi.org/10.1289/EHP6960

Hu, Z., Lee, J.W., Chandran, K., Kim, S., & Khanal, S.K. (2012). Nitrous oxide (N_2O) emission from aquaculture: A review. *Environmental Science & Technology, 46*, 6470–6480. https://doi.org/10.1021/es300110x

Hussain, K.A., Romanova, S., Okur, I., Zhang, J., & Keubler, J. (2023). Assessing the release of microplastics and nanoplastics plastic containers and reusable food pouches: Implications for human health. *Environmental Science & Technology, 57*, 9782–9792.

International Agency for Research on Cancer. (2015, March 20). *Monographs volume 112: Evaluation of five organophosphate insecticides and herbicides.* World Health Organization. www.iarc.who.int/wp-content/uploads/2018/07/MonographVolume112-1.pdf

Jamieson, A.J., Brooks, L.S.R., Reid, W.D.K., Piertney, S.B., Narayanaswamy, B.E., & Linley, T.D. (2019). Microplastics and synthetic particles ingested by deep-sea amphipods in six of the deepest marine ecosystems on Earth. *Royal Society Open Science, 6*(180667), 2–11. http://dx.doi.org/10.1098/rsos.180667

Klaassen, M., & Willie, M. (2023). Wild birds' plight and role in the current bird flu panzootic. *bioRxviv*. https://doi.org/10.1101/2023.05.02.539182

Kuiken, T., & Cromie, R. (2022). Protect wildlife from livestock diseases. *Science, 378*(6615), 5–6. https://doi.org/10.1126/science.adf0956

Landrigan, P.J., Raps, H., Cropper, M., Bald, C., Brunner, M., Canonizado, E.M., Charles, D., Chiles, T.C., Donohue, M.J., Enck, J., Fenichel, P., Fleming, L.E., Ferrier-Pages, C., Fordham, R., Gozt, A., Griffin, C., Hahn, M.E., Haryanto, B., Hixson, R., Ianelli, H., … & Dunlop, S. (2023). The Minderoo-Monaco commission on plastics and human health. *Annals of Global Health*, 89(1), 1–215. https://doi.org/10.5334/aogh.4056

Law, K.L., Starr, N., Siegler, T.R., Jambeck, J.R., Mallos, N.J., & Leonard, G.H. (2020). The United States' contribution of plastic waste to land and ocean. *Science Advances*, *6*(eabd0288), 1–7.

Liu, C.M., Aziz, M., Park, D.E., Wu, Z., Stegger, M., Li, M., Wang, Y., Schmidlin, K., Johnson, T.J., Koch, B.J., Hungate, B.S., Nordstrom, L., Gauld, L., Weaver, B., Rolland, D., Statham, S., Hall, B., Sariya, S., Davis, G.S., … & Price, L. (2023). Using source-associated mobile genetic elements to identify zoonotic extraintestinal *E. coli* infections. *One Health*, *16*(100518), 1–8. https://doi.org/10.1016/j.onehlt.2023.100518

Manges, A.R., Smith, S.P., Lau, B.J., Nuval, C.J., Eisenberg, J.N.S., Dietrich, P.S., & Riley, L.W. (2007). Retail meat consumption and the acquisition of antimicrobial resistant *Escherichia coli* causing urinary tract infections: A case–control study. *Foodborne Pathogens and Disease*, *4*(4), 419–431.

Miller, S.A., Ferreira, J.P., & LeJeune, J.T. (2022). Antimicrobial use and resistance in plant agriculture: A one health perspective. *Agriculture*, *12*(289), 1–27. https://doi.org/10.3390/agriculture12020289

Morales-Caselles, C. (2021). An inshore–offshore sorting system revealed from global classification of ocean litter. *Nature Sustainability*, *4*, 484–493. https://doi.org/10.1038/s41893-021-00720-8

Mulder, M., Kiefte-de Jong, J.C., Goessens, W.H.F., de Visser, H., Ikram, M.A., Verbon, A., & Stricker, B.H. (2019). Diet as a risk factor for antimicrobial resistance in community-acquired urinary tract infections in a middle-aged and elderly population: A case-control study. *Clinical Microbiology and Infection*, *25*, 613–619. https://doi.org/10.1016/j.cmi.2018.07.029

National Academies of Sciences, Engineering, and Medicine [NAS]. (2022). *Reckoning with the U.S. role in global ocean plastic waste.* The National Academies Press. https://doi.org/10.17226/26132.

Nelson, R.E., Hatfield, K.M., Wolford, H., Samore, M.H., Scott, R.D., Reddy, R.C., Olubajo, B., Paul, P., Jernigan, J.A., & Baggs, J. (2021). National estimates of healthcare costs associated with multidrug- resistant bacterial infections among hospitalized patients in the United States. *Clinical Infectious Diseases, 72*(Suppl 1), S17–S26. https://doi.org/10.1093/cid/ciaa1581

Panieri, E., Baralic, K., Djukic-Cosic, D., Djordjevic, A.B., & Saso, L. (2022). PFAS molecules: A major concern for the human health and the environment. *Toxics*, *10*(44), 1–55. https://doi.org/10.3390/toxics10020044

Perrone, A. (2019 August 18). Unprecedented heatwave 'kills thousands of fish' in Alaska. *The Independent.* www.the-independent.com/climate-change/news/alaska-heatwave-salmon-rivers-july-temperatures-climate-change a9063461.html#:~:text= Climate %20 change%20and%20warming%20 rivers,gripped%20the%20state%20last%20month.

Puryear, W., Sawatzki, K., Hill, N., Foss, A., Stone, J.J., Doughty, L., Walk, D., Gilbert, K., Murray, M., Cox, E., Patel, P., Mertz, Z., Ellis, S., Taylor, J., Fauquier, D., Smith, A., DiGiovanni, R.A., van de Guchte, A., Gonzalez-Reiche, A.S., … & Runstadler, J. (2023). Highly pathogenic avian influenza A(H5N1) virus outbreak in New England seals, United States. *Emerging Infectious Diseases*, *29*(4), 786–791.

Reverter, M., Sarter, S., Caruso, D., Avarre, J.C., Combe, M., Pepey, E., Pouyaud, L., Vega-Heredia, S., de Verdal, H., & Gozlan, R.E. (2020). Aquaculture at the crossroads of global warming and antimicrobial resistance. *Nature Communications*, *11*(1870), 1–8. https://doi.org/10.1038/s41467-020-15735-6

Richardson, K., Steffen, W., Lucht, W., Bendtsen, J., Cornell, S.E., Donges, J.F., Drüke, M., Fetzer, I., Bala, G., von Bloh, W., Feulner, G., Fiedler, S., Gerten, D., Gleeson, T. Hofmann, M., Huiskamp, W., Kummu, M., Mohan, C., Nogués-Bravo1, D., … & Rockström, J. (2023). Earth beyond six of nine planetary boundaries. *Science Advances*, *9*, 1–16.

Rizan, C., Mortimer, F., Stancliffe, R., & Bhutta, M.F. (2020). Plastics in healthcare: Time for a re-evaluation. *Journal of the Royal Society of Medicine*, *113*(2), 49–53. https://doi.org/10.1177/0141076819890554

Rochman, C.M., Brookson, C., Bikker, J., Djuric, N., Earn, A., Bucci, K., Athey, S., Huntington, A., McIlwraith, H., Munno, K., De Frond, H., Kolomijeca, A., Erdle, L., Grbic, J., Bayoumi, M., Borrelle, S.B., Wu, T., Santoro, S., Werbowski, L.M., … & Hung, C. (2019). Rethinking microplastics as a diverse contaminant suite. *Environmental Toxicology and Chemistry*, *38*(4), 703–7111.

Royer, S.J., Ferron, S., Wilson, S.T., & Karl, D.M. (2018). Production of methane and ethylene from plastic in the environment. *PLoS ONE*, *13*(8), 1–13. https://doi.org/10.1371/journal.pone.0200574

Schar, D., Zhao, C., Wang, Y., Joakim Larsson, D.G., Gilbert, M., & Van Boeckel, T.P. (2021). Twenty-year trends in antimicrobial resistance from aquaculture and fisheries in Asia. *Nature Communications*, *12*(5384), 1–10. https://doi.org/10.1038/s41467-021-25655-8

Schwartz-Narbonne, H., Xia, C., Shalin, A., Whitehead, H.D., Yang, D., Peaslee, G.F., Wang, Z., Wu, Y., Peng, H., Blum, A., Venier, M., & Diamond, M.L. (2023). Per- and polyfluoroalkyl substances in Canadian fast food packaging. *Environmental Science & Technology Letters*, *10*, 343–349. https://doi.org/10.1021/acs.estlett.2c00926

Shepon, A., Wu, T., Kremen, C., Dayan, T., Perfecto, I., Fanzo, J., Eshel, G., & Golden, C.D. (2023). Exploring scenarios for the food system–zoonotic risk interface. *Lancet Planetary Health*, *7*, e329–35.

Skandalis, N., Maeusli, M., Papafotis, D., Miller, S., Lee, B., Theologidis, I., & Luna, B. (2021). Environmental spread of antibiotic resistance. *Antibiotics*, *10*(640), 1–14. https://doi.org/10.3390/antibiotics10060640

Steffen, W., Rockström, J., Richardson, K., Lenton, T.M., Folke, C., Liverman, D., Summerhayes, C.P., Barnosky, A.D., Cornella, S.E., Crucifixi, M., Donges, J.F., Fetzera, I., Lade, S.J., Marten Scheffer, M., Winkelmann, R., & Schellnhuber, H.J. (2018). Trajectories of the earth system in the Anthropocene. *Proceedings of the National Academy of Sciences*, *115*(33), 8252–8259.

Tang, K.L., Caffrey, N.P., Nobrega, D.B., Cork, S.C., Ronksley, P.E., Barkema, H.W., Polachek, A.J., Ganshorn, H., Sharma, N., Kellner, J.D. & Ghali, W.A. (2017). Restricting the use of antibiotics in food-producing animals and its associations with antibiotic resistance in food-producing animals and human beings: a systematic review and meta-analysis. *Lancet Planetary Health*, 1, e316-327. http://dx.doi.org/10.1016/S2542-5196(17)30141-9

Thorton, P., Nelson, G., Mayberry, D., & Herrero, M. (2022). Impacts of heat stress on global cattle production during the 21st century: A modelling study. *Lancet Planetary Health*, *6*, e192–201.

Tian, H., Xu, R., Canadell, J.G., Thompson, R.L., Winiwarter, W., Suntharalingam, P., Davidson, E.A., Ciais, P., Jackson, R.B., Janssens-Maenhout, G., Prather, M.J., Regnier, P., Pan, N., Pan, S., Peters, G.P., Shi, H., Tubiello, F.N., Zaehle, S., Zhou, F., … & Yao, Y. (2020). A comprehensive quantification of global nitrous oxide sources and sinks. *Nature*, *586*, 248–256. doi:10.1038/s41586-020-2780-0

Tudi, M., Ruan, H.D., Wang, L., Lyu, J., Sadler, R., Connell, D., Chu, C., & Phung, D.T. (2021). Agriculture development, pesticide application and its impact on the environment. *International Journal of Environmental Research and Public Health*, *18*(1112), 1–24. https://doi.org/10.3390/ ijerph18031112

Tuvo, B., Scarpaci, M., Bracaloni, S., Esposito, E., Costa, A.L., Ioppolo, M., & Casini, B. (2023). Microplastics and antibiotic resistance: The magnitude of the problem and the emerging role of hospital wastewater. *International Journal of Environmental Research and Public Health*, *20*(5868), 1–14. https://doi.org/10.3390/ijerph20105868

United Nations Environment Programme & Secretariat of the Basel, Rotterdam and Stockholm Conventions. (2023). *Chemicals in plastics: A technical report.* UNEP. www.unep.org/resources/report/chemicals-plastics-technical-report

United Nations Environment Assembly of the United Nations Environment Programme. (2022). *Resolution adopted by the United Nations Environment Assembly on 2 March 2022. 5/14 End plastic pollution: Towards an international legally binding instrument.* UNEP United Nations Environment Program.. https://digitallibrary.un.org/record/3999257?ln=en&v=pdf

United Nations Environment Programme. (2022). *Plastics in agriculture – an environmental challenge. Foresight Brief 029.* United Nations Environment Program UNEP. https://wedocs.unep.org/bitstream/handle/20.500.11822/40403/Plastics_Agriculture.pdf

United Nations Environment Programme & International Livestock Research Institute (2020). *Preventing the next pandemic: Zoonotic diseases and how to break the chain of transmission.* www.unep.org/resources/report/preventing-future-zoonotic-disease-outbreaks-protecting-environment-animals-and

U.S. Environmental Protection Agency. (2020). *Advancing sustainable materials management: 2018 fact sheet.* www.epa.gov/sites/default/files/2020-11/documents/2018_ff_fact_sheet.pdf

U.S. Geological Survey [USGS]. (2019). *Agricultural contaminants.* www.usgs.gov/mission-areas/water-resources/science/agricultural-contaminants

Van Boeckel, T.P., Brower, C., Gilbert, M., Grenfell, B.T., Levin, S.A., Robinson, T.P., Teillant, A., & Laxminarayan, R. (2015). Global trends in antimicrobial use in food animals. *Proceedings of the National Academy of Sciences, 112*(18), 5649–5654. www.pnas.org/cgi/doi/10.1073/pnas.1503141112

Wallinga, D. (2021, November 18). U.S. livestock antibiotic use is rising, medical use falls. *National Resources Defense Council.* www.nrdc.org/experts/david-wallinga-md/us-livestock-antibiotic-use-rising-medical-use-falls-0

Wallinga, D., Smit, L.A.M., Davis, M.F., Casey, J.A., & Nachman, K.E. (2022). A review of the effectiveness of current US policies on antimicrobial use in meat and poultry production. *Current Environmental Health Reports*, *9*, 339–354. https://doi.org/10.1007/s40572-022-00351-x

Watts, J.E.M., Schreier, H.J., Lanska, L., & Hale, M.S. (2017). The rising tide of antimicrobial resistance in aquaculture: Sources, sinks and solutions. *Marine Drugs*, *15*(158), 1–16. https://doi.org/10.3390/md15060158

West, J.J., Fiore, A.M., Horowitz, L.W., & Mauzerall, D.L. (2006). Global health benefits of mitigating ozone pollution with methane emission controls. *Proceedings of the National Academy of Sciences*, *103*(11), 3988–3993. https://doi.org10.1073 pnas.0600201103

World Economic Forum, Ellen MacArthur Foundation and McKinsey & Company (2016). *The new plastics economy — Rethinking the future of plastics.* https://emf.thirdlight.com/file/24/_ABkCs_skP18I_Am1g_JWxFrX/The%20New%20Plastics%20Economy%3A%20Rethinking%20the%20future%20of%20plastics.pdf

World Health Organization. (2017, November 7). *Stop using antibiotics in healthy to animals prevent the spread of antibiotic resistance.* [Press release]. www.who.int/news/item/07-11-2017-stop-using-antibiotics-in-healthy-animals-to-prevent-the-spread-of antibioticresistance#:~:text=WHO%20is%20recommending%20that%20farmers,their%20unnecessary%20use%20in%20animals.

World Health Organization. (2021). *Global antimicrobial resistance and use surveillance system (GLASS) report.* www.who.int/publications/i/item/9789240062702

World Health Organization. (2023). *Antimicrobial resistance.* www.who.int/news-room/fact-sheets/detail/antimicrobial-resistance

World Meteorological Organization. (2025, January 10). *WMO confirms 2024 as warmest year on record at about 1.55°C above pre-industrial level.* https://wmo.int/news/media-centre/wmo-confirms-2024-warmest-year-record-about-155degc-above-pre-industrial-level#:~:text=pre%2Dindustrial%20level-,WMO%20confirms%202024%20as%20warmest%20year%20on%20record%20at%20about,C%20above%20pre%2Dindustrial%20level&text=The%20World%20Meteorological%20Organization%20(WMO,based%20on%20six%20international%20datasets.

Yang, C.Z., Yaniger, S.I., Jordan, V.C., Klein, D.J., & Bittner, G.D. (2011). Most plastic products release estrogenic chemicals: A potential health problem that can be solved. *Environmental Health Perspectives*, *119*(7), 989–996.

Zahm, S., Bonde, J.P., Chiu, W.A., Hoppin, J., Kanno, J., Abdallah, M., Dong, G.H., Dorman, D.C., Fry, R., Guo, H., Haug, L.S., Hoffman, J.N., Iwasaki, M., Machala, M., Mancini, F.R., Maria-Engler, S.S., Moller, P., Ng, J.C., Pallardy, M., … & Woskie, S. (2023). Carcinogenicity of perfluorooctanoic acid and perfluorooctanesulfonic acid. *Lancet Oncology*, *21*(1), 16–17. https://doi.org/10.1016/S1470-2045(23)00622-8

Zhang, L., Rana, I., Shaffer, R.M., Taioli, E., & Sheppard, L. (2019). Exposure to glyphosate-based herbicides and risk for Non-Hodgkin Lymphoma: A meta-analysis and supporting evidence. *Mutation Research*, *78*, 186–206. https://doi.org/10.1016/j.mrrev.2019.02.001

Zhu, C., Kobayashi, K., Loladze, I., Zhu, J., Jiang, Q., Xu, X., Liu, G., Seneweera, S., Ebi, K.L., Drewnowski, A., Fukagawa, N.K., & Ziska, L.H. (2018). Carbon dioxide (CO2) levels this century will alter the protein, micronutrients, and vitamin content of rice grains with potential health consequences for the poorest rice-dependent countries. *Science Advances*, *4*, 1–8.

Zimmerman, L., Dierkes, G., Ternes, T.A., Volker, C., & Wagner, M. (2019). Benchmarking the in vitro toxicity and chemical composition of plastic consumer products. *Environmental Science & Technology*, *53*, 11467–11477. https://doi.org/10.1021/acs.est.9b02293

Zimmerman, L., Dombrowski, A., Volker, C., & Wagner, M. (2020). Are bioplastics and plant-based materials safer than conventional plastics? In vitro toxicity and chemical composition. *Environment International*, *145*, 1–11. https://doi.org/10.1016/j.envint.2020.106066

The Global Burden of Diet-Related Chronic Disease

5

"We know enough to prevent and control most CVD [cardiovascular disease]; however, knowing is not enough."

George A. Mensah[1], MD, Valentin Fuster[1], MD, PhD & Gregory Roth[1], MD, MPH

OVERVIEW

Global food consumption has shifted from homemade fresh foods to the increased consumption of animal products and processed foods while undernutrition, hunger, and food insecurity impacts billions across the world (Popkin, 2005; Food and Agriculture Organization [FAO] et al., 2023). The global triple burden of malnutrition (stunting, wasting, and overweight) impacting children under five years of age is immense: over 148 million children suffered from stunting in the year 2022; 45 million were threatened with wasting; and 37 million children were obese, respectively (United Nations Children's Fund [UNICEF] et al., 2023). Historically, infectious diseases claimed the most lives across the globe but chronic diseases (also known as noncommunicable diseases/NCDs) are now responsible for over 70% of all deaths globally (World Health Organization [WHO], 2023a). Low and middle-income countries bear the burden of 86% of premature deaths due to NCDs (WHO, 2023a). Seven of the top ten causes of death globally were NCDs in 2019 with diet playing a major role in many

DOI: 10.1201/9781003153528-6

(WHO, 2020). As nutrition, lifestyle behaviors, and health equity play a prominent role in NCDs, this chapter will largely focus on the global burden of diet and chronic disease amongst adults. It will also highlight the role of whole plant foods nutrition in chronic disease prevention.

THE GLOBAL BURDEN OF DISEASE AND THE DIETARY PATTERNS OF ADULTS

The 2017 Global Burden of Disease, Injuries, and Risk Factors Study team of diet collaborators analyzed data from 195 countries to understand the impact of 15 dietary factors on adult NCD morbidity, mortality, and socioeconomic development at the population level (Global Burden Disease 2017 Diet Collaborators [GBDDC], 2019). Suboptimal dietary intake was responsible for more deaths amongst adults than any other risk factors including smoking where dietary risk was associated with 22% of all deaths among adults and 15% of all disability-adjusted life years/DALY (GBDDC, 2019). The acronym DALY is an epidemiological measurement that represents the number of years of life lost due to premature mortality and the number of years lived with a disability from the respective disease or condition (GBDDC, 2019; WHO, n.d.). Cardiovascular disease dominated the largest percentage of deaths (GBDDC, 2019).

This study also found that adults across the globe consume suboptimal amounts of nutrient-dense plant foods such as nuts, seeds, and whole grains while overconsuming sugar-sweetened drinks, red meat, processed meats, and sodium (GBDDC, 2019). The leading diet-related causes of death came from diets high in sodium and diets low in whole grains, fruits, vegetables, nuts, seeds, and omega-3 fatty acids. High sodium intake, low grain intake, and low fruit intake were responsible for greater than one-half of diet-related deaths and two-thirds of DALY (GBDDC, 2019). Increasing healthy dietary habits could prevent at least 20% of deaths globally (GBDDC, 2019).

Utilizing Global Burden of Disease (GBD) data from 1990 to 2019, a ~32% increase in deaths globally were attributed to low fruit intake (Pan et al., 2023). A diet low in fruit equates to consuming less than 250 grams per day (approximately one cup/8 ounces). The countries associated with the highest mortality due to low fruit intake were India, China, and Russia, respectively (Pan et al., 2023). Fruits are an important source of antioxidants, polyphenols, phytonutrients, and fiber (Dreher, 2018). Fiber is an invaluable component of a healthy dietary pattern and is a "food component of public health concern"; intakes are low in the U.S. and around the globe (Dietary Guidelines Advisory

Committee, 2020; Zhuo et al., 2022). The data underscores the critical importance of whole plant foods nutrition in disease prevention.

The Burden of High Red Meat and Processed Meat Consumption

From 1990–2019, ischemic heart disease death (the blockage of coronary arteries that can cause a heart attack) due to high red meat consumption (18-27 grams per day of beef, lamb, pork, or goat) increased as did deaths due to high processed meat consumption (0-4 grams per day of cured, smoked, salted, or chemically preserved meat) (Yan et al., 2023). China had the highest number of deaths from high red meat intake followed by the U.S. and then Russia, respectively (Yan et al., 2023). The U.S. had the highest burden of ischemic heart disease due to high processed meat consumption followed by Russia and India, respectively (Yan et al., 2023). Additionally, Wu et al. (2023) assessed global deaths and DALYs attributed to high red meat consumption (defined as more than 23 grams per day of beef, pork, lamb, or goat) from 1999–2019 and found a steady increase in mortality and DALYs.

A well-publicized review of the literature by Lescinsky et al. (2022) assessed the role of unprocessed meat and chronic disease by utilizing a "Burden of Proof Risk Function" methodology; their results defied a large body of evidence that preceded it as they found weak evidence of harmful associations between unprocessed red meats and the risk of colorectal cancer, breast cancer, ischemic heart disease, Type 2 diabetes, ischemic stroke, and hemorrhagic stroke (Lescinsky et al., 2022). These researchers did state however, "…we contend that consuming no unprocessed red meat likely minimizes the risk of health consequences compared to consuming any…" (Lescinsky et al., 2022). A critical analysis of this study and methodology found that it was flawed and unjustifiable (Glenn et al., 2023).

Cardiovascular Disease

The burden of cardiovascular disease is immense. Utilizing GBD data from 1990–2019, Roth et al. (2020) found that the global prevalence of cardiovascular disease (CVD) doubled from 1990 to 2019; cardiovascular deaths increased during that time too (Roth et al., 2020). The role of diet cannot be overstated in CVD where transgressions in dietary intake is attributed to death and DALY (Roth et al., 2020). In 2019, diet and lack of physical activity were amongst the top modifiable risk factors for cardiovascular disease (Roth et al., 2020). Low levels of physical activity are problematic globally and attributed to increased mortality, DALY, and years of life lost (YLL) (Roth et al., 2020). The prevalence of stroke increased during this time period, too (Roth et al., 2020).

Ischemic heart disease is responsible for a large portion of cardiovascular disease (Wang et al., 2021). Wang et al. (2021) analyzed GBD data between 1990–2019 to better understand the role of metabolic risk factors and ischemic heart disease. They found that the burden of ischemic heart disease has shifted from countries of higher sociodemographic status to those of lower sociodemographic status (Wang et al., 2021). In regions of higher sociodemographic status and well-controlled metabolic risk factors, the risk of mortality from ischemic heart disease decreases (Wang et al., 2021). Premature death from ischemic heart disease was more prominent in developing regions where low to middle-income regions experienced increases across all age groups (Wang et al., 2021). This not only demonstrates the need for effective large-scale interventions in regions that are becoming increasingly urbanized—but points to health disparities amongst vulnerable populations.

Importantly, whole plant foods nutrition is now recognized as an asset in improving outcomes in persons with cardiovascular disease where healthy plant-based diets are associated with decreased cardiovascular disease risk mortality and all-cause mortality (Kim et al., 2019). Esselstyn et al. (2014) studied adherence to and the outcomes of 198 self-selected patients with cardiovascular risk factors that followed a dietary regimen exclusively of whole plant foods nutrition that was oil-free; exercise was encouraged but not mandated. In this cohort, 89% of patients complied with the regimen where the clinical data indicated that cardiovascular disease progression was largely stopped, reversed, or halted (Esselstyn et al., 2014). These researchers concluded that the prevailing, mainstream approach to cardiovascular medicine "…can neither cure the disease nor end the epidemic and is financially unsustainable" (Esselstyn et al., 2014).

High Blood Pressure and Hyperlipidemia

Globally, persons with high systolic blood pressure/HSBP increased from over 2 billion people to over 4 billion from 1990–2019 (Roth et al., 2020). High blood pressure (known as "the silent killer") is a risk factor for heart disease. The 2023 World Health Organization report on hypertension draws attention to nearly 11 million avoidable premature deaths globally and 235 million DALY annually (WHO, 2023b).

Elevated low-density lipoprotein-C (LDL-C) levels known as "bad" cholesterol increased globally from 1990–2019 along with the associated total number of deaths and DALY (Roth et al., 2020; Zheng et al., 2022). The burden of high LDL-C increasingly impacts low-income countries (Zheng et al., 2022).

Cancer

The burden of cancers has increased globally, with a ~26% increase in the number of new cases and a nearly 21% increase in deaths from 1990–2019 (Global Burden of Disease 2019 Cancer Collaboration, 2022). There was also a 16% increase in cancer DALY, with persons in low and middle-income quintiles experiencing the largest percent increase in cancer cases and deaths (Global Burden of Disease 2019 Cancer Collaboration, 2022). The total number of cancer deaths amongst females globally in 2019 came from breast cancer followed by lung cancer, colorectal cancers, and stomach cancer, respectively (Global Burden of Disease 2019 Cancer Collaboration, 2022). For men, lung cancer accounted for the most deaths followed by stomach cancer, colorectal cancer, and prostate cancer, respectively (Global Burden of Disease 2019 Cancer Collaboration, 2022). Over 44% of cancer deaths were attributed to modifiable risk factors (Global Burden of Disease 2019 Cancer Collaboration, 2022).

The International Agency for Research on Cancer (IARC) determined that red meat (goat, mutton, lamb, horse, beef, veal, pork) *is probably carcinogenic* to humans while processed meats (any meat that has been salted, cured, fermented, smoked, or preserved) *is carcinogenic* to humans (International Agency for Research on Cancer [IARC], 2018). More specifically, positive evidence and mechanistic associations were found between processed meats and colorectal and stomach cancers; positive associations and mechanistic evidence was found between red meat and colorectal, pancreatic, and prostate cancers (Bouvard et al., 2015).

Prediabetes

Prediabetes is a condition in which glucose tolerance or fasting glucose is impaired which increases the risk of Type 2 Diabetes (T2D) (Rooney et al., 2023). Prediabetes is considered a serious health condition which is projected to increase globally (; Rooney et al., 2023). In the U.S., more than one in three persons is living with pre-diabetes (Centers for Disease Control and Prevention [CDC], 2024). Globally, 9% of adults ages 20-79 have increased glucose tolerance levels and nearly 6% had impaired fasting glucose levels (Rooney et al., 2023). People living in low-income countries are projected to have a marked increase in both impaired glucose tolerance and impaired fasting glucose compared to middle and high-income countries, highlighting a glaring health disparity (Rooney et al., 2023).

Diabetes

Diabetes is a critical public health threat where in 2021, ~529 million people across the globe were living with diabetes of which 96% of cases were attributed to T2D (Global Burden of Disease 2021 Diabetes Collaborators, 2023). Although T2D is largely preventable with healthy lifestyle behaviors, it's projected that 1.3 billion people will be living with diabetes by 2050 (Global Burden of Disease 2021 Diabetes Collaborators, 2023). In 2021, high Body Mass Index (BMI) accounted for ~53% of T2D DALY (Global Burden of Disease 2021 Diabetes Collaborators, 2023). Diabetes is a risk factor for other serious conditions such as cardiovascular disease and chronic kidney disease (CDC, 2022).

Additional research sheds light on the global burden of diet-related T2D (O'Hearn et al., 2023). From 1990–2018, researchers found that 70% (~14 million) of new cases of T2D were attributed to diet (O'Hearn et al., 2023). Globally, the largest proportion of the total diet-attributable disease burden came from refined rice, refined wheat, processed meats, unprocessed red meat, sugar-sweetened beverages, excessive potato intake, and fruit juices (O'Hearn et al., 2023). Individually, insufficient whole grain intake along with excessive intakes of refined rice, refined wheat, processed meat, and unprocessed red meat were linked to the burden of disease (O'Hearn et al., 2023). Another global analysis found that without effective interventions, the number of deaths, incidence, DALY, and prevalence of diabetes will increase globally between the year 2018 to 2025 (Lin et al., 2020). In low-income countries, diets low in whole grains, nuts, seeds, and fruits were the leading dietary risk factors for T2D (Lin et al., 2020).

A scoping review of the literature to determine connections between diabetes (Type 1 & Type 2), climate change/global warming, and environmental racism found that people living with diabetes in communities already marginalized by environmental racism were increasingly vulnerable to the negative impacts of climate change; diabetes morbidity increased with increased temperatures; extreme heat was a risk factor for mortality; prolonged heat stress was associated with higher excess mortality; and air pollutants impacted physiological outcomes (Ospelt et al., 2023).

Obesity

Considered a pandemic, obesity is a risk factor for many chronic conditions such as heart disease, T2D, and certain cancers (CDC, 2022b). In 2017, 2.4 million deaths among women and 2.3 million deaths among men globally were attributed to high BMI, respectively (BMI > 25 kg weight/height m^2)

(Dai et al., 2020). Cardiovascular disease was the leading cause of high-BMI disability-adjusted life years, which was followed by diabetes and kidney disease (Dai et al., 2020). Over 800 million people across the globe are obese and it's estimated that over a billion people will be obese by 2030, including persons in low-income countries (WHO, 2023c).

The circumstances surrounding obesity are complex which includes ubiquitous obesogenic food environments that involve a multitude of forces across systems, industries, institutions, and policies (WHO, 2023c; Walker et al., 2020). Obesogenic food environments are dynamic and are associated with but are not limited to:

1) The vast amounts of money spent on food and social marketing to both adults and children (American Heart Association, 2019; Nieto et al., 2022).
2) Food costs and agricultural subsidies; pricing strategies in the U.S. could reduce diet-related NCD deaths and health disparities (Peñalvo et al., 2017). Importantly, creating global agricultural subsidies for nutritious foods that includes whole plant foods nutrition would increase fruit and vegetable consumption and reduce diet-related mortality all while improving planetary health outcomes (Springman & Freund, 2022).
3) Lack of proper healthcare infrastructure, structural racism, health inequities—and the role of the built environment (lack of greenspace, environmental racism, and residential segregation) (Thorpe et al., 2017; Bell et al., 2019; Churchwell et al., 2020; De la Fuente et al., 2021; Devakumar et al., 2022; WHO, 2022; Hill et al., 2023). These issues will be discerned in forthcoming chapters.
4) Shifts in dietary patterns including increased animal product intake and the consumption of ultra-processed and fast foods (Bodirsky et al., 2020; Coppola et al., 2023; Fazzino et al., 2023; Hall et al., 2019; Wang & Beydoun, 2009; Wang et al., 2022; You & Henneberg, 2016).
5) Ingestion/exposure to antibiotics and chemicals including plastics and agrichemicals as well as aberrations to the gut microbiome (Amato et al., 2021; Egusquiza 2020; Heindel et al., 2024; Jiang et al., 2022; Kladnicka et al., 2022; Peresellin & Tempkin, 2022; Ren et al., 2020; Kannan & Vimalkumar, 2021; Liu et al., 2021; Vallianou et al., 2021;)
6) Physical inactivity, inadequate sleep, increased screen time, and the limited amount of time spent cooking—as well as meals eaten outside the home (American Public Health Association [APHA], 2021; CDC, 2022; Chaput et al., 2023; Haghjoo et al., 2022; Kolodinsky & Goldstein, 2011; Landais et al., 2022; Mills et al., 2017; Silveira et al., 2022).

Reducing the prevalence of overweight and obesity by 5% would offer a global savings of $429 billion per year; the cost of obesity (globally) weighs in at $990 billion per year (WHO, 2023c). By 2025, the cost of treating complications associated with obesity will reach $1 trillion dollars globally (WHO, 2023c).

Metabolic Syndrome

Lastly, metabolic syndrome (also called insulin resistance syndrome) is a name given to a largely preventable cluster of conditions that increase the risk of chronic diseases such as cardiovascular disease, stroke, diabetes, etc. (National Heart, Lung, & Blood Institute [NHLBI], 2022). It's characterized by abdominal obesity, high blood sugar levels, high blood lipid levels and low high-density lipoprotein (HDL) (NHLBI, 2022). One in three Americans has metabolic syndrome (NHLBI, 2022). Xing et al. (2023) found high BMI and high fasting blood glucose were important contributors to the burden of metabolism-related neoplasms, too.

NOTE

1 Mensah, G.M., et al. (2023). A heart-healthy and stroke-free world. *Journal of the American College of Cardiology, 82*(25), 2343–2349. https://doi.org/10.1016/j.jacc.2023.11.003

REFERENCES

Amato, A.A., Wheeler, H.B., & Blumberg, B. (2021). Obesity and endocrine-disrupting chemicals. *Endocrine Connections*, *10*, R87–R105. https://doi.org/10.1530/EC-20-0578

American Heart Association. (2019, April 9). *Fact sheet: Unhealthy and unregulated food advertising and marketing to children.* www.heart.org/en/-/media/Files/About-Us/Policy-Research/Fact-Sheets/Healthy-Schools-and-Childhood-Obesity/Food-Marketing-and-Advertising-to-Children-Fact-Sheet.PDF?sc_lang=en

American Public Health Association. (2021, October 26). *Supporting physical education in schools for all youth. Policy Number: 20211.* https://apha.org/Policies-and-Advocacy/Public-Health-Policy-Statement…ase/2022/01/07/Supporting-Physical-Education-in-Schools-for-All-Youth

Bell, C.N, Kerr, J., & Young, J.L. (2019). Associations between obesity, obesogenic environments, and structural racism vary by county-level racial composition.

International Journal of Environmental Research and Public Health, 16(861), 1–17. https://doi.org/10.3390/ijerph16050861

Bodirsky, B.L., Dietrich, J.P., Martinelli, E., Stenstad, A., Pradhan, P., Gabrysch, S., Mishral, A., Weindl, I., Mouë, C.L., Rolinski, S., Baumstark, L., Wang, X., Waid, J.L., Lotze-Campen, H., & Popp, A. (2020). The ongoing nutrition transition thwarts long-term targets for food security, public health and environmental protection. *Scientific Reports, 10*(19778), 1–14. https://doi.org/10.1038/s41598-020-75213-3

Bouvard, V., Loomis, D., Guyton, K.Z., Grosse, Y., Ghissassi, F.E., Benbrahim-Tallaa, L., Guha, N., Mattock, H., & Straif, K. (2015). Carcinogenicity of consumption of red and processed meat. *Lancet Oncology, 16*, 1599–1600.

Centers for Disease Control and Prevention. (2024, May 15). *1 in 3 Americans have prediabetes.* www.cdc.gov/diabetes/communication-resources/1-in-3-americans.html

Centers for Disease Control and Prevention. (2022a, December 30). *Diabetes and chronic kidney disease.* www.cdc.gov/diabetes/managing/diabetes-kidney-disease.ht…iseases%20are%20the%209th,adults%20with%20diabetes%20has%20CKD.

Centers for Disease Control and Prevention. (2022b, May 17). *Adult obesity facts.* www.cdc.gov/obesity/data/adult.html#:~:text=Obesity%2Drela…ditions%20include%20heart,%24173%20billion%20in%202019%20dollars.

Chaput, J.P., McHill, A., Cox, R.C., Broussard, J.L., Dutil, C., da Costa, B.G.G., Sampasa-Kanyingal, H., & Wright, K.P. (2023). The role of insufficient sleep and circadian misalignment in obesity. *Nature Reviews Endocrinology, 19*, 82–97. https://doi.org/10.1038/s41574-022-00747-7

Churchwell, K., Elkind, M., Benjamin, R., Carson, E.K., Lawrence, W., Mills, A., Odom, T.M., Rodriguez, F., Sanchez, E., Sharrief, A.Z., Sims, M., & Williams, O. (2020). Call to action: Structural racism as a fundamental driver of health disparities. A presidential advisory from the American Heart Association. *Circulation, 142*, e454–e468. https://doi.org/10.1161/CIR.0000000000000936

Coppola, S., Lorella Paparo, L., Trinchese, G., Andrea Margarita Rivieri, A.M., Masino, A., Severina, A.F., Cerulo, M., Escolino, M., & Turco, A., (2023). Increased dietary intake of ultraprocessed foods and mitochondrial metabolism alterations in pediatric obesity. *Scientific Reports, 13*(12609), 1–9. https://doi.org/10.1038/s41598-023-39566-9

Dai, H., Alsalhe, T.A., Chalghaf, N., Nicola, M.R., Bragazzi, L., & Wu, J. (2020). The global burden of disease attributable to high body mass index in 195 countries and territories, 1990–2017: An analysis of the Global Burden of Disease Study. *PLoS Med, 17*(7), e1003198. https://doi.org/10.1371/journal.pmed.1003198

De la Fuente, F., Saldías, M.A., Cubillos, C., Mery, M., Carvajal, D., Bowen, M., & Bertoglia, M.P. (2021). Green space exposure association with Type 2 Diabetes Mellitus, physical activity, and obesity: A systematic review. International Journal of Environmental Research and Public Health, *18*(97), 1–18. https://dx.doi.org/10.3390/ijerph18010097

Devakumar, D., Selvarajah, S., Abubakar, I., Kim, S.S., McKee, M., Sabharwal, N.S., Saini, A., Shannon, G., White, A.I.R., & Achiume, E.T. (2022). Racism, xenophobia, discrimination, and the determination of Health. *Lancet, 400*, 2097–2108.

Dietary Guidelines Advisory Committee. (2020). *Scientific report of the 2020 dietary guidelines advisory committee: Advisory report to the secretary of agriculture and the secretary of health and human services*. U.S. Department of Agriculture, Agricultural Research Service. https://doi.org/10.52570/DGAC2020

Dreher, M.L. (2018). Whole fruits and fruit fiber emerging health effects. *Nutrients, 10*(1833), 1–54. https://doi.org/10.3390/nu10121833

Egusquiza, R.J., & Blumberg, B. (2020). Susceptibility to obesity: New mechanisms and chemicals. *Endocrinology, 161*(3), 1–14. https://doi.org/10.1210/endocr/bqaa024_

Esselstyn, C.B., Gendy, G., Doyle, J., Golubic, M., & Roizen, M.F. (2014). A way to reverse CAD? *Journal of Family Practice, 63*(7), 356–364.

Fazzino, T.L., Junn, D., Chollet-Hinton, L., & Bjorlie, K. (2023). US tobacco companies selectively disseminated hyper-palatable foods into the US food system: Empirical evidence and current implications. *Addiction, 119*(1),1–10. https://doi.org/10.1111/add.16332

Food and Agriculture Organization [FAO], International Fund for Agricultural Development [IFAD], United Nations Children's Fund [UNICEF], World Food Programme [WFP] & World Health Organization [WHO]. (2023). The State of Food Security and Nutrition in the World 2023. Urbanization, agrifood systems transformation and healthy diets across the rural–urban continuum. https://doi.org/10.4060/cc3017en

Glenn, A., Gu, X., Hu, F.B., Wang, M., & Willett, W.C. (2023). Concerns about the burden of proof studies. *Nature Medicine, 29*, 823–825. https://doi.org/10.1038/s41591-023-02294-8

Global Burden of Disease 2021 Diabetes Collaborators. (2023). Global, regional, and national burden of diabetes from 1990 to 2021, with projections of prevalence to 2050: A systematic analysis for the Global Burden of Disease Study 2021. *Lancet, 402*, 203–34. https://doi.org/10.1016/S0140-6736(23)01301-6

Global Burden Disease 2017 Diet Collaborators. (2019). Health effects of dietary risks in 195 countries, 1990–2017: A systematic analysis for the Global Burden of Disease Study 2017. *Lancet, 393*, 1958–1972.

Global Burden of Disease 2019 Cancer Collaboration. (2022). Cancer incidence, mortality, years of life lost, years lived with disability, and disability-adjusted life years for 29 cancer groups from 2010 to 2019. *JAMA Oncology, 8*(3), 420–444. https://doi.org/10.1001/jamaoncol.2021.6987

Haghjoo, P., Siri, G., Soleimani, E., Farhangi, M.A., & Alesaeidi, S. (2022). Screen time increases overweight and obesity risk among adolescents: A systematic review anddose-response meta-analysis. *BMC Primary Care, 23*(161), 1–24. https://doi.org/10.1186/s12875-022-01761-4

International Agency for Research on Cancer [IARC]. (2018). Red meat and processed meat, volume 114. IARC monographs on the evaluation of carcinogenic risks to humans. www.ncbi.nlm.nih.gov/books/NBK507971/

Jiang, F., Peng, Y.E., & Sun, Q. (2022). Pesticides exposure induced obesity and its associated diseases: Recent progress and challenges. *Journal of Future Foods,* 2(2), 119–124. https://doi.org/10.1016/j.jfutfo.2022.03.005

Hall, K.D., Ayuketah, A., Brychta, R., Walter, P.J., Yang, S., & Zhou, M. (2019). Ultra-processed diets cause excess calorie intake and weight gain: An inpatient randomized controlled trial of ad libitum food intake. *Cell Metabolism, 30*, 67–77. https://doi.org/10.1016/j.cmet.2019.05.008

Heindel, J.J., Lustig, R.H., Howard, S., & Corkey, B.E. (2024). Obesogens: A unifying theory for the global rise in obesity. *International Journal of Obesity, 48*, 449–460. https://doi.org/10.1038/s41366-024-01460-3

Hill, L., Nambi, N., & Artiga, S. (2023, March 29). *Key data on health and health care by race and ethnicity*. Kaiser Family Foundation. www.kff.org/racial-equity-and-health-policy/report/key-data-on-health-and-health-care-by-race-and-ethnicity/view/print/

Kannan, K., & Vimalkumar, K. (2021). Review of human exposure to microplastics and insights into microplastics as obesogens. *Frontiers in Endocrinology, 12*(724989), 1–19. https://doi.org.10.3389/fendo.2021.724989

Kim, H. Caulfeld, L.E., Garcia-Larsen, V., Steffen, L.M., Coresh, J., & Rebholz, C.M. (2019). Plant-based diets are associated with a lower risk of incident cardiovascular disease, cardiovascular disease mortality, and all-cause mortality in a general population of middle-aged adults. *Journal of the American Heart Association, 8*(e012865), 1–13. https://doi.org/10.1161/JAHA.119.012865

Kladnicka, I., Bludovska, M., Plavinova, I., Muller, L., & Mullerova, D. (2022). Obesogens in foods. *Biomolecules, 12*(680), 1–18. https://doi.org/10.3390/biom12050680

Kolodinsky, J.M., & Goldstein, A.B. (2011). Time use and food pattern influences on obesity. *Obesity, 19*(12), 2327–2335. https://doi.org/10.1038/oby.2011.130

Landais, E., Miotto-Plessis, M., Bene, C., d'Hotel, E.M., Truong, M.T., Some, J.W., & Verger, E.O. (2022). Consumption of food away from home in low-and middle-income countries: A systematic scoping review. *Nutrition Reviews, 81*(6), 727–754. https://doi.org/10.1093/nutrit/nuac085

Lescinsky, H., Afshin, A., Ashbaugh, C. Bisignano, C., Brauer, M., Ferrara, G., Hay, S.I., He, J., Iannucci, V., Marczak, L.B., McLaughlin, S.A., Mullany, E.C., Parent, M.C., Serfes, A.L., Sorensen, R.J.D., Aravkin, A.Y., Zheng, P., & Murray, C.J.L. (2022). Health effects associated with consumption of unprocessed red meat: A burden of proof study. *Nature Medicine, 28*, 2075–2082. https://doi.org/10.1038/s41591-022-01968-z

Lin, X., Xu, Y., Pan, X., Xu, J., Ding, Y., Sun, X., Song, X., Ren, Y., & Shan, P. (2020). Global, regional, and national burden and trend of diabetes in 195 countries and territories: An analysis from 1990 to 2025. *Scientific Reports, 10*(14790), 1–11. https://doi.org/10.1038/s41598-020-71908-9

Liu B.N., Liu, X.T., Liang, Z.H., & Wang, J. (2021). Gut microbiota in obesity. *World Journal of Gastroenterology, 27*(25), 3837–3850. https://doi.org/10.3748/wjg.v27.i25.3837

Mills, S., Brown, H., Wrieden, W., White, M., & Adams, J. (2017). Frequency of eating home cooked meals and potential benefits for diet and health: Cross-sectional analysis of a population-based cohort study. *International Journal of Behavioral Nutrition and Physical Activity, 14*(109), 1–11. https://doi.org/10.1186/s12966-017-0567-y

National Heart, Lung & Blood Institute (2022, May 18). *What is metabolic syndrome?* National Institutes of Health. www.nhlbi.nih.gov/health/metabolic-syndrome

Nieto, C., Jáuregui, A., Contreras-Manzano, A., Kent, M.P., Sacks, G., White, C.M., Pauzé, E., Vanderlee, L., Thrasher, J., Barquera, S., & Hammond, D. (2022). Adults' exposure to unhealthy food and beverage marketing: A multi-country study in Australia, Canada, Mexico, the United Kingdom, and the United States. *Journal of Nutrition, 152*, 25S–34S. https://doi.org/10.1093/jn/nxab449.

O'Hearn, M., Lara-Castor, L., Cudhea, F., Miller, V., Reedy, J., Shi, P., Zhang, J., Wong, J.B., Economos, C.D., Michal, R., Mozaffarian, D., & Global Dietary Database. (2023). Incident type 2 diabetes attributable to suboptimal diet in 184 countries. *Nature Medicine*, *29*, 982–995. https://doi.org/10.1038/s41591-023-02278-8

Ospelt, E., Hardison, H., Mungmode, A., Noor, N., Bankole, K., Anikpezie, N., Chima, C., Singh, A.K., Wright, T., Elisha, V., Shah, V.N., & Ebekozien, O. (2023). The impact of climate change on people living with diabetes: A scoping review. *Clinical Diabetology*, *12*(3), 1–15.

Pan, S., Lin, Z., Yao, T., Guo, X., Xu, T., Sheng, X., Song, X., Chen, Z., Wei, W., Yan, Y., & Hu, Y. (2023). Global burden of non-communicable chronic diseases associated with a diet low in fruits from 1990 to 2019. *Frontiers in Nutrition*, *10*(1202763), 1–12. https://doi.org/10.3389/fnut.2023.1202763

Peresellin, K., & Tempkin, A. (2022). *Chemicals in our food may be contributing to weight gain.* Environmental Working Group. www.ewg.org/news-insights/news/2022/09/chemicals-our-food-may-be-contributing-weight-gain

Peñalvo, J.L., Cudhea, F., Micha, R., Rehm, C.D., Afshin, A., Whitsel, L., Wilde, P., Gaziano, T., Pearson-Stuttard, J., O'Flaherty, M., Capewell, S., & Mozaffarian, D. (2017). The potential impact of food taxes and subsidies on cardiovascular disease and diabetes burden and disparities in the United States. *BMC Medicine*, *15*(208), 1–13. https://doi.org/10.1186/s12916-017-0971-9

Popkin, B.M. (2005). Using research on the obesity pandemic as a guide to a unified vision of nutrition. *Public Health Nutrition*, *8*(6A), 724–729. https://doi.org/10.1079/PHN2005776

Ren, X.M., Kuo, Y., & Blumberg, B. (2020). Agrochemicals and obesity. *Molecular and Cellular Endocrinology*, *515*(110926), 1–38. https://doi.org/10.1016/j.mce.2020.110926

Rooney, M.R., Fang, M., Ogurtsova, K., Ozkan, B., Echouffo-Tcheugui, J.B., Boyko, E.J., ... & Selvin, E. (2023). Global prevalence of prediabetes. *Diabetes Care*, *46*(7), 1388–1394. https://doi.org/10.2337/dc22-2376

Roth, G.A, Mensah, G.A., Johnson, C.O., Addolorato, G., Ammirati, E., Baddour, L.M., Barengo, N.C., Beaton, A.Z., Benjamin, E.J., Benziger, C.P., Bonny, A., Brauer, M., Brodmann, M., Cahill, T.J., Carapetis, J., Catapano, A.L., Chugh, S.S., Cooper, L.T., Coresh, J., … & Furster, V. (2020). Global burden of cardiovascular diseases and risk factors, 1990–2019. *Journal of the American College of Cardiology*, *76*(25), 2982–3021. https://doi.org/10.1016/j.jacc.2020.11.010

Silveira, E.A., Mendonca, C.R., Delpino, F.M., Souza, G.V.E., de Souza Rosa, L.P., de Oliveira, C., & Noll, M. (2022). Sedentary behavior, physical inactivity, abdominal obesity and Obesity in adults and older adults: A systematic review and meta-analysis. *Clinical Nutrition ESPEN, 50*(63e73), 63–73. https://doi.org/10.1016/j.clnesp.2022.06.001

Springmann, M., & Freund, F. (2022). Options for reforming agricultural subsidies from health, climate, and economic perspectives. *Nature Communications*, *13*(82), 1–7. https://doi.org/10.1038/s41467-021-27645-2

Thorpe, R.J., Parker, L.J., Cobb, R.J., Dillard, F., & Bowie, J. (2017). Association between discrimination and obesity in African-American men. *Biodemography and Social Biology, 63*(3), 253–261. https://doi.org/10.1080/19485565.2017.1353406

United Nations Children's Fund [UNICEF], World Health Organization [WHO], International Bank for Reconstruction and Development/The World Bank. (2023). *Levels and trends in child malnutrition: UNICEF/WHO /World Bank Group joint child malnutrition estimates: Key findings of the 2023 edition.* UNICEF and WHO. https://iris.who.int/bitstream/handle/10665/368038/9789240073791-eng.pdf?sequence=1

Vallianou, N., Dalamaga, M., Stratigou, T., Karampela, I., & Tsigalou, C. (2021). Do antibiotics cause obesity through long-term alterations in the gut microbiome? A Review of current evidence. *Current Obesity Reports*, *10*, 244–262. https://doi.org/10.1007/s13679-021-00438-w

Walker, B.B., Shashank, A., Gasevic, D., Schuurman, N., Poirier, P., Teo, K., Rangarajan, S., Yusuf, S., & Lear, S.A. (2020). The local food environment and obesity: Evidence from three cities. *Obesity*, *28*, 40–45.

Wang, W., Hu, M., Liu, H., Yang, H., Cai, J., & Li, H. (2021). Global Burden of Disease Study 2019 suggests that metabolic risk factors are the leading drivers of the burden of ischemic heart disease. *Cell Metabolism*, *33*, 1943–1956. https://doi.org/10.1016/j.cmet.2021.08.005

Wang, Y., & Beydoun, M.A. (2009). Meat consumption is associated with obesity and central obesity among US adults. *International Journal of Obesity*, *33*(6), 621–628. https://doi.org/10.1038/ijo.2009.45

Wang, Y., Wang, K., Du, M., Khandpur, N., Rossato, S.L., Lo, C., VanEvery, H., Kim, D.Y., Zhang, F.F., Chavarro, J.E., Sun, Q., Huttenhower, C., Song, M., Nguyen, L.H., & Chan, A.T. (2022). Maternal consumption of ultra-processed foods and subsequent risk of offspring overweight or obesity: Results from three prospective cohort studies. *BMJ*, *379*(e071767), 1–12. http://dx.doi.org/10.1136/bmj2022-071767

World Health Organization. (n.d.). *Disability-adjusted life.* www.who.int/data/gho/indicator-metadata-registry/imr-detai…20equivalent%20of%20one,health%20condition%20in%20a%20population.

World Health Organization. (2020, December 10). *The top 10 causes of death.* www.who.int/news-room/fact-sheets/detail/the-top-10-causes-of-death

World Health Organization. (2022). *Strengthening primary health care to tackle racial discrimination, promote intercultural services and reduce health inequities: Research brief.* https://iris.who.int/bitstream/handle/10665/363854/9789240057104-eng.pdf?sequence=1

World Health Organization. (2023a, September 16). *Noncommunicable diseases.* www.who.int/news-room/fact-sheets/detail/noncommunicable-diseases

World Health Organization. (2023b). *Global report on hypertension: The race against a silent killer.* www.who.int/publications/i/item/9789240081062

World Health Organization. (2023c). *WHO acceleration plan to stop obesity*. www.google.com/books/edition/WHO_acceleration_plan_to_stop_obesity/86IOEQAAQBAJ?hl=en&gbpv=1&printsec=frontcover

Wu, Z., Wang, W., Zhang, K., Fan, M., & Lin, R. (2023). The global burden of disease attributable to diet high in red meat in 204 countries and territories, 1999–2019: An updated analysis of the Global Burden of Disease Study. *Molecular Nutrition & Food Research*, *67*(2300144), 1–10. https://doi.org/10.1002/mnfr.20230014

Xing, Q.Q., Li, J.M., Z.J. Chen, Lin, XY, You, YY, Hong, M.Z., Weng, S., & Pan, J.S. (2023). Global burden of common cancers attributable to metabolic risks from 1990-2019. *Med*, 4, 168–181.

Yan, D., Liu, K., Li, F., Shi, D., Wei, L., Zhang, J., ... & Wang, Z. (2023). Global burden of ischemic heart disease associated with high red and processed meat consumption: An analysis of 204 countries and territories between 1990 and 2019. *BMC Public Health*, *23*(2267), 1–15. https://doi.org/10.1186/s12889-023-16954-4

You, W., & Henneberg, M. (2016). Meat consumption providing a surplus energy in modern diet contributes to obesity prevalence: An ecological analysis. *BMC Nutrition*, *2*(22), 1–11. https://doi.org/10.1186/s40795-016-0063-9

Zheng, J., Wang, J., Zhang, Y., Xia, J., Guo, H., Hu, H., Shan, P., & Li, T. (2022). The global burden of diseases attributed to high low-density lipoprotein cholesterol from 1990 to 2019. *Frontiers in Public Health*, *10*(891929), 1–16. https://doi.org/10.3389/fpubh.2022.891929

Zhuo, M., Chen, Z., Zhong, M., Liu, Y., Lei, F., Qin, J., Sun, T., Yang, C., Chen, M., Song, X., Wang, L., Li1, Y., Zhang, X., Zhu, L., Cai, J., Ye, J., Zhou, G., & Zeng, Y. (2022). The global disease burden attributable to a diet low in fibre in 204 countries and territories from 1990 to 2019. *Public Health Nutrition*, *26*(4), 854–865. https://doi.org/10.1017/S1368980022001987

Structural Violence, Environmental Racism and the Social Determinants of Health

6

"Not all people are polluted equally…"

Daniel Faber[1], PhD, Benjamin Levy[1], PhD & Christina Schlegel[1], MUPP

OVERVIEW

Before delving into the health disparities associated with nutrition-related chronic disease in the U.S. in Chapter 7, this chapter offers a brief summary of the tentacles of racism that has shaped the Social Determinants of Health (SDH), health inequities, and other broadscale injustices that are forms of structural violence. These injustices include but are not limited to food-source disparities, forced labor, environmental pollution, concentrated animal feeding

 DOI: 10.1201/9781003153528-7

operations (CAFO), global extractivism, and ecocide. The chapter concludes by highlighting the principles of food sovereignty (Nyelni Food Sovereignty Steering Committee, 2007).

STRUCTURAL RACISM AND STRUCTURAL VIOLENCE IN THE U.S.

Structural racism in the U.S. stems from the history of colonialism, slavery, the takeover of Indigenous land, segregation, and sanctions which have imparted negative outcomes for Indigenous persons and persons of color, immigrants, those impoverished, and women (Bailey et al., 2021; Bravemen et al., 2022; Javidan 2021; Kaiser Family Foundation, n.d.; Sharif et al., 2022). These persons can be described as "Historically Oppressed and other Peoples Experiencing Inequities" (HOPEI) (American Public Health Association [APHA], 2022). Both overt and clandestine biases have impacted policies, institutions, employment, health outcomes, unequal access to healthcare, education and housing, redlining, disproportionate incarceration, generational wealth, and the unequal distribution of resources for HOPEI (Bailey et al., 2021; Freidrichs, 2011; KFF, n.d.; Kozol, 1992; National Association for the Advancement of Colored People [NAACP], 2025; Prison Policy Institute, n.d.).

Structural racism can be defined as ideas and situations that go beyond individually held prejudices to impact laws, rules, and norms that become embedded throughout the constructed culture, the economic system, and the policies that shape how people live (Bailey et al., 2021). As an example, persons of color, with the exception of Asian Americans, are imprisoned at disproportional rates compared to white Americans in the U.S. (Prison Policy Initiative, n.d.). Notably, the U.S. has the highest rate of incarceration in the world (Widra & Herring, 2021). Racism can be defined as, "…structures, policies, practices, and norms that assigns value and determines opportunity based on the way people look or the color of their skin" (Centers for Disease Control and Prevention [CDC], 2023). The CDC declares racism "a serious threat to public health" (CDC, 2023). As of August of 2021, there were 209 declarations by towns, cities, health associations, public health departments, county boards, governors, etc. in 37 states in the U.S. that proclaimed racism is a public health crisis (APHA, 2021).

The APHA calls structural racism a form of structural violence (APHA, 2022; CDC, 2023). The APHA defines structural violence as, "…a form of oppression in which a given social structure harms people by preventing them from meeting their basic needs and being valued" (APHA, 2022). Sharif et al. (2022) have conveyed that structural racism and structural violence are interconnected threats to health, creating health inequities. Such oppression leads to social factors that influence and undermine health and well-being, called the Social Determinants of Health (SDH).

THE SOCIAL DETERMINANTS OF HEALTH, HEALTH DISPARITIES, AND INEQUALITY

The SDH are the conditions that shape our lives and our health outcomes and includes where we live, how we live, employment security, food security, housing security, early childhood experiences, access to healthcare, social inclusion, etc. (World Health Organization [WHO], n.d.). *Healthy People 2030*, the nation's blueprint for improved population health goals, categorizes the SDH into five domains as noted in Figure 6.1 (U.S. Department of Health and Human Services [US DHHS], n.d.).

The APHA (n.d.) describes racial equity as, "…the condition that will be achieved when racial identity no longer predicts how a person fares in society". Racial equity is central to health equity that "…allows everyone to achieve their best health" (APHA, n.d.). The WHO defines health equity as "…the absence of unfair and avoidable or remediable differences in health among population groups defined socially, economically, demographically or geographically" (WHO, n.d.). Similar to the WHO definition, *Healthy People 2030* defined health equity as individuals able to attain their highest level of health when everyone is valued equally—and avoidable injustices and inequalities are remedied (Ndugga & Artiga, 2023). Additionally, The Health Equity Institute (n.d.) defines health equity as "efforts to ensure that all people have full and equal access to opportunities that enable them to lead healthy lives". Systems have long been set up to reward some and ignore others as social systems "are not naturally inequitable" (The George Washington University, 2021). Justice prevails when inequities are addressed and systemic barriers are eliminated (Thomas, 2023).

A health disparity can be defined as "a particular type of health difference that is closely linked with social, economic, and/or environmental disadvantage." (US DHHS, 2022). McCartney et al. (2019) offers an overarching definition for health inequalities, "Health inequalities are the systematic, avoidable and unfair differences in health outcomes that can be observed between populations, between social groups within the same population or as a gradient across a population ranked by social position."

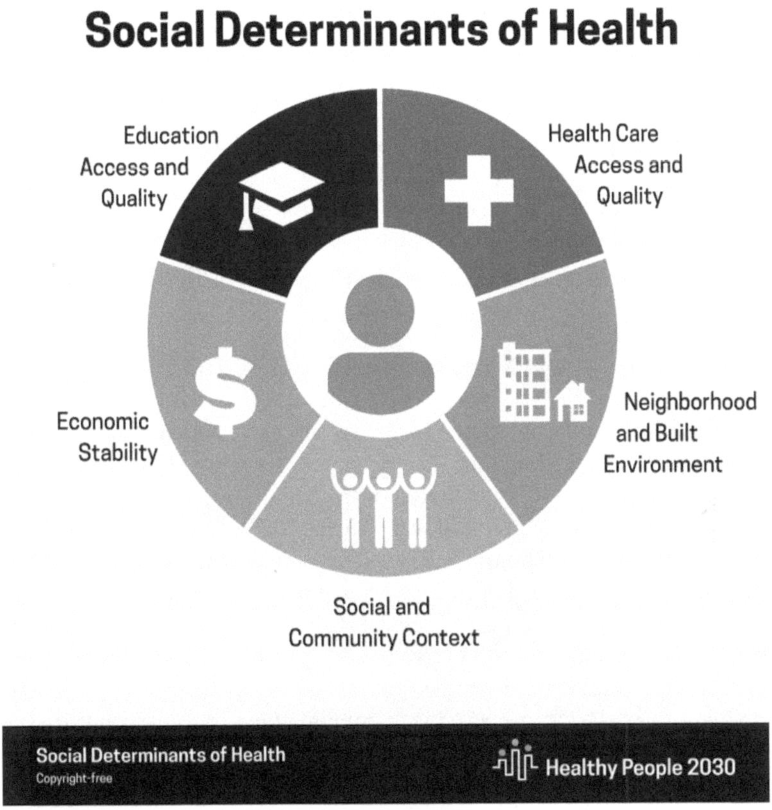

FIGURE 6.1 Healthy People 2030, Social Determinants of Health, U.S. Department of Health and Human Services.

ENVIRONMENTAL RACISM

Environmental racism is a form of systemic racism where a business/industry creates pollution that increases the risk of and/or creates negative outcomes for humans and the natural environment; low-income regions and HOPEI are largely targeted as norms and policies have allowed these injustices to prevail (Ihejirika, 2023). In 1980, the U.S. Congress created the Comprehensive Environmental Response, Compensation, and Liability Act (CERCLA) also known as "Superfund" to clean up toxic waste sites and hold polluting businesses/commercial industries accountable—as well as authorize monies for cleanup under certain circumstances (U.S. EPA, 2023a). People of color, Indigenous persons, and those with low-incomes are more likely to live within a 3-mile radius of a Superfund site than the rest of the population (U.S. EPA, 2023b).

"Brownfields" are a property, development, or expansion where its respective use is hampered by the presence or potential presence of hazardous contaminants, pollutants, or substances (U.S. EPA, 2016). There are more than 450,000 Brownfields in the U.S. (U.S. EPA, 2024). Between 2015–2019, 149 million people in the U.S. lived within a 3-mile radius of a Brownfield which included 55% of all households below the poverty line, 59% of all blacks, 52% of all Hispanics, 46% of all children in the U.S. under the age of 5, and 51% of all people with less than a high school education, respectively (U.S. EPA & Office of Land and Emergency Management, 2021).

Cancer Alley and Corporate Pollution

An egregious example of environmental racism comes from "Cancer Alley", a name given to the Mississippi River Chemical Corridor in Louisiana (U.S.) that's situated on a ~130 mile stretch of the Mississippi River between New Orleans and Baton Rouge (Terrell & St Julien, 2022). Data analyzed between 2008–2017 to assess relationships (if at all) between cancer incidence, toxic air assessment, census tracts, tumors, etc., found that industrial air pollution significantly drives new cancer cases in this region where black and impoverished communities are under assault (Terrell & St. Julien, 2022). Previously, James et al. (2012) examined cancer risk and air toxins in Cancer Alley in 2005 and found that low-income and black communities were disproportionately impacted where several toxic compounds exceeded benchmark levels and cancer risk was significantly higher (James et al., 2012). They also found that cancer risk increased with decreasing household income; cancer risk also increased with a greater percentage of Black residents in a community (James et al., 2012).

Living within close proximity to Louisiana's Denka Performance Elastomer Plant (formerly owned by DuPont and now the only manufacturer of neoprene in the United States) is significantly associated with cancer prevalence where cancer rates are 71% higher than the national average (University Network for Human Rights, 2019). These massive industrial complexes have also transformed crucial ecosystem services and their invaluable first line defense from catastrophic climate events such as mitigating the impact of storms and severe flooding (National Oceanic and Atmospheric Administration [NOAA], 2024a). In fact, Louisiana's rate of wetland loss is amongst the highest in the world (NOAA, 2024b).

Polluting industries are not limited to imposing negative outcomes on HOPEI communities. Without admitting guilt, the 3M Company settled a $850 million-dollar (USD) lawsuit with the state of Minnesota (MN) in 2018 for environmental contamination with forever chemicals (Minnesota Pollution Control Agency, 2025; Winter, 2022). The impact of their groundwater contamination in the largely White suburban city of Oakdale, MN has been devastating; there was a 171% increased likelihood of childhood cancer, a significant decline in the general fertility rate, and an increased likelihood of premature or low birth weight babies (Sunding, 2017). Cancer-associated mortality for adults was greater in Oakdale compared to communities outside the contamination area, too (Sunding, 2017). The U.S. Environmental Protection Agency (EPA) reports that the chemical manufacturing industry in the U.S. generates the most waste amongst industries—even more than petroleum and coal products manufacturing (Yavrom, 2021).

Businesses that plunder the living systems of natural environment and negatively impact human health are sometimes thought of as "necessary" because they provide jobs for those in the impacted area. Messenger's (2024a) study of the Louisiana Industrial Tax Exemption Program found that tax subsidies offered no significant relationship to local job growth or personal income growth. Instead, these massive industrial tax subsidies created harmful outcomes to local residents (Messenger, 2024a). Additionally, a study of the Shell Chemical Appalachia Polymers plastic pellets refinery in Pennsylvania found that a flawed economic analysis projected unrealistic local economic benefits that have not been realized despite billions of dollars in tax payer subsidies—all while the plastics refinery engaged in environmental violations (Messenger et al., 2024b).

Global Environmental Racism

Environmental racism is evident across the globe (Achiume, 2022; Navas et al., 2022). Some multinational corporations have entrenched histories in controlling and profiting from the natural resources of less-developed countries despite the negative externalities imposed on local citizens and the natural environment (Achiume, 2019). This global "extractivism economy" can be defined as, "…the industries, actors and financial flows, as well as the economic, material and social processes and outputs, associated with the globalized extraction of natural resources" (Achiume, 2019). These extractivism economies are linked to racial and environmental injustices around the world that stem from colonialism, slavery, the extirpation of Indigenous persons, and systematic oppression (Achiume, 2019; Achiume, 2022; Hardoon, 2017).

Industrial Agriculture, Forced Labor, and Environmental Racism

Agriculture transects racism, environmental racism, and environmental degradation "…through labor exploitations, toxin and carcinogen exposure, access to food, and agricultural run-off and waste" (O'Donnell, 2024). Forced labor is a form of modern slavery where escaping the situation is highly improbable due to coercion, psychological abuse, violence, etc. (International Labour Organization et al., 2022). Substandard and dangerous working and housing conditions (or inaccessibility to decent housing) are prevalent in the food industry which includes but is not limited to poultry, cattle, and dairy production—and agrichemical exposure (Hall et al., 2021; Oxfam America, 2015; Environmental Justice Foundation, 2023; Donley et al., 2022; Sanchez & Jameel, 2024).

Blackstone et al. (2023) analyzed forced labor risk amidst food supply chains and global trade patterns. Meat, poultry, and eggs most contributed to forced labor risk (28%) followed by coffee, beer, wine, cocoa, etc. (23%)—and then processed vegetables (18%), respectively. They also found a disproportionately high risk of forced labor was associated with boneless beef, dried milk and cheese—as well as cocoa and refined sugars. The fruits that disproportionately increased risk were avocados, lemons, limes and pineapples while vegetables included tomatoes, chilies, and peppers, respectively (Blackstone et al., 2023).

Concentrated Animal Feeding Operations and Environmental Racism

Concentrated Animal Feeding Operations (CAFO) describes the conditions where vast numbers of animals are housed, fed, and confined for the purposes of fast growth and development (US. EPA, n.d.). The environment impact of CAFO's is well documented and includes but is not limited to: diminished air quality; dead zones, water, nutrient, and chemical pollution; groundwater contamination; antibiotic resistant genes and pathogenic microorganisms; and soil contamination (Petitioners to EPA, 2022). Communities of color and disenfranchised communities are disproportionately impacted by industrial animal agriculture (Ayashi & Elsheikh, 2015; Donley et al., 2022; Hall et al., 2021; Son et al., 2021).

NATIONAL AND GLOBAL FOOD-SOURCE DISPARITIES

Lack of access to healthy foods is another manifestation of structural violence. Food security can be defined as, "access by all people at all times to enough food for an active, healthy life" whereas food insecurity can be defined as the inability to acquire adequate foods due to a lack of money or resources (Rabbit et al., 2023). Black and Hispanic households experienced rates of food insecurity that were significantly higher than the national average in 2022 where 16% of Black households and over 13% of Hispanic households included food insecure children compared to 5.5% of white households, respectively (Rabbitt et al., 2023). Food insecurity impacted nearly 9% of U.S. households with children in 2022, accounting for over 3 million people (Rabbit et al., 2023).

A "food desert" has been defined as an "area in the United States with limited access to affordable and nutritious food, particularly such an area composed of predominantly lower income neighborhoods and communities" (United States Department Apartment of Agriculture, 2009). In 2012, Dutko et al. reported that persons living in census tracts with food deserts included higher percentages of people of color; people experiencing higher rates of unemployment; those with lower incomes; and areas with an increased proportion of vacant and abandoned buildings compared to census tracts that had lower percentages of people of color. However, some studies in the U.S. have demonstrated that access to grocery stores in underserved areas is not always associated with increased food access or increased fruit and vegetable consumption (Cummins et al., 2014; Elbel et al., 2015). "Food swamps" are characterized by readily accessible, highly processed foods that are less expensive and can negatively impact health outcomes (Frayne & McCordic, 2018).

In the U.S., substandard food environments are linked to higher rates of obesity, obesity-related cancer, and kidney disease amongst Black individuals (Bevel et al., 2023; Cerceo et al., 2023; Ferrara et al., 2024). Those experiencing severe food insecurity are more likely to have a chronic disease (Gregory & Coleman-Jensen, 2017). Analyzing NHANES data from 2011–2015, Gregory and Coleman-Jensen (2017) found that decreased food security was associated with a higher probability of chronic diseases such as kidney disease, hypertension, heart disease, stroke, hepatitis, cancer, asthma, diabetes, arthritis, and chronic obstructive pulmonary disease.

Globally, nearly 30% of the world's population were moderately or severely food insecure, where 2.4 billion people lack access to sufficient, safe, and nutritious food during the year (Food and Agriculture Organization [FAO] et al., 2023). Over 40% of the global population cannot afford a healthy diet (Food and Agriculture Organization [FAO] et al., 2023). These problems are compounded by the insufficient availability of nutritious foods along with the increased marketing of processed foods (Food and Agriculture Organization [FAO] et al., 2023). Fraval et al. (2019) assessed over 6,000 rural land-holding households in sub-Saharan Africa where 39% were considered severely food insecure. One-half of children impacted by stunting live in Asia; 43% live in Africa (United Nations Children's Fund [UNICEF] et al., 2023). Additionally, over 2 billion people globally lack access to safely managed drinking water while over 3.5 billion people do not have access to safely managed sanitation (United Nations, n.d.).

FOOD SOVEREIGNTY

The right to food is a universal human right where nations are called upon to ensure food security while responsibly protecting the natural environment and agricultural biodiversity (United Nations Human Rights Council, 2019). Food sovereignty includes but is not limited to, "…the right of peoples to healthy and culturally appropriate food produced through ecologically sound and sustainable methods, and their right to define their own food and agriculture systems. It puts those who produce, distribute and consume food at the heart of food systems and policies rather than the demands of markets and corporations" (Declaration of Nyleni, 2007). Food sovereignty also ensures care of the environment for future generations where food and food providers are valued—and food builds knowledge and skills in unison with nature (The Nyéléni International Steering Committee, 2007).

In closing, structural violence is manifested in a myriad of ways throughout the world including the predominant economic structures, policies, and food systems.

NOTE

1 Faber, D., Levy, B., & Schlegel, C. (2021). Not all people are polluted equally in capitalist society: An eco-socialist commentary on liberal environmental justice theory. *Capitalism Nature Socialism*, *32*(4), 1–16. DOI: 10.1080/10455752.2021.2009640

REFERENCES

Achiume, T. (2019, May 14). *Global extractivism and racial equality*. United Nations General Assembly. www.ohchr.org/en/documents/thematic-reports/ahrc4154-global-extractivism-and-racial-equality-report-special

Achiume, T. (2022, October 25). *Report of the Special Rapporteur on contemporary forms of racism, racial discrimination, xenophobia and related intolerance on ecological crisis climate justice and racial justice.* United Nations General Assembly. https://promiseinstitute.law.ucla.edu/wpcontent/uploads/2022/10/A_77_2990_AdvanceUneditedVersion.pdf

American Public Health Association. (n.d.). Racial equity & public health. www.apha.org//media/files/pdf/advocacy/speak/210825_racial_equity_fact_sheet.ashx

American Public Health Association. (2021, October). *Analysis: Declarations of racism as a public health crisis.* www.apha.org/-/media/files/pdf/topics/racism/racism_declarations_analysis.ashx

American Public Health Association. (2022, November 8). *A strategy to address systemic racism and violence as public health priorities: Training and supporting community health workers to advance equity and violence prevention.* Policy Number 20227. www.apha.org/policies-and-advocacy/public-health-policy-statements/policy-database/2023/01/18/address-systemic-racism-and-violence

Ayashi, H., & Elsheikh, E. (2015). *The U.S. Farm Bill: Corporate power and structural racialization. in the United States food system.* The Haas Institute for a Fair and Inclusive Society, University California Berkeley. https://escholarship.org/uc/item/55v6q06x

Bailey, Z.D., Feldman, J.M., & Basset, M.T. (2021). How structural racism works — Racist policies as a root cause of. U.S. Racial Health Inequities. *New England Journal of Medicine*, *384*(8), 768–773.

Bevel, M.S., Tsai, M., Parham, A., Andrzejak, S.A., Jones, S., & Moore, J.X. (2023). Association of food deserts and food swamps with obesity-related cancer mortality in the US. *JAMA Oncology*, *9*(7), 909–916. https://doi.org/10.1001/jamaoncol.2023.0634

Blackstone, N.T., Rodriguez-Huerta, E., Battaglia, K., Jackson, B., Jackson, E., Norris, C.B. & Sparks, J.L.D. (2023). Forced labour risk is pervasive in the US land-based food supply. *Nature Food*, 4, 596-606.

Bravemen, P.A., Arkin, E., Proctor, D., Kuah, T., & Holm, N. (2022). Systemic and structural racism: Definitions, examples, health damages, and approaches to dismantling. *Health Affairs*, *41*(2), 171–178. https://doi.org/10.1377/hlthaff.2021.01394

Centers for Disease Control and Prevention [CDC]. (2023, September 18). *Racism and health.* www.cdc.gov/minorityhealth/racism-disparities/index.html

Cerceo, E., Sharma, E., Boguslavsky, A., & Rachoin, J.S. (2023). Impact of food environments on obesity rates: A state-level analysis. *Journal of Obesity*, 2023, Article ID 5052613, 1–7. https://doi.org/10.1155/2023/5052613

Cummins, S., Flint, E., & Matthews, S.A. (2014). New neighborhood grocery store increased awareness of food access but did not alter dietary habits or obesity. *Health Affairs, 33*(2), 283–291.

Donley, N., Bullard, R.D., Economos, J., Figueroa, I., Lee, J., Liebman, A.K., Martinez, D.N., & Shafie, F. (2022). Pesticides and environmental injustice in the USA: root causes, current regulatory reinforcement and a path forward. *BMC Public Health*, *22*(708), 1–23. https://doi.org/10.1186/s12889-022-13057-4

Dutko, P., Ver Ploeg, M., & Farrigan, T. (2012, August). *Characteristics and influential factors of food deserts.* ERR-140. U.S. Department of Agriculture,

Economic Research Service. https://ers.usda.gov/sites/default/files/_laserfiche/publications/45014/30940_err140.pdf?v=48431

Environmental Justice Foundation. (2023, December). *Slave labour in the Brazilian cattle ranching industry: The case of the Pantanal and the European market.* EJF. https://ejfoundation.org/resources/downloads/EJF-EU-Pantanal-human-rights-report-cattle-sector-2023-v5.pdf

Ferrara, F., Siligato, R., Di Maria, A., Scichilone, L., Di Simone, E., Bondanelli, M., Storari, A., De Giorgi, A., Di Muzio, M.D., & Fabbian, B. (2024). Food insecurity and kidney disease: A systematic review. *International Urology and Nephrology, 56*, 1035–1044. https://doi.org/10.1007/s11255-023-03777-w

Food and Agriculture Organization [FAO], International Fund for Agricultural Development [IFAD], United Nations Children's Fund [UNICEF], World

Food Programme [WFP] & World Health Organization [WHO]. (2023). The state of food security and nutrition in the world 2023. Urbanization, agrifood systems transformation and healthy diets across the rural–urban continuum. https://doi.org/10.4060/cc3017en

Fraval, S., Hammond, J., Bogard, J.R., Ng'endo, M., van Etten, J., Herrero, M., Oosting, S.J., de Boer, I.J.M., Lannerstad, M., Teufel, N., Lamanna, C., Rosenstock, T.S., Pagella, T., Vanlauwe, B., Dontsop-Nguezet, P.M., Baines, D., Carpena, P., Njingulula, P., Okafor, C…van Wijk, M.T. Food access deficiencies in Sub-saharan Africa: Prevalence and implications for agricultural interventions. (2019). *Frontiers in Sustainable Food Systems, 3*(104), 1-13. doi: 10.3389/fsufs.2019.00104

Frayne, B., & McCordic, C. (2018). Food swamps and poor dietary diversity: Longwave development implications in southern African cities. *Sustainability, 10*(4425), 1–12. https://doi.org/10.3390/su10124425

Freidrichs, C. (Director). (2011). *The Pruitt-Igoe myth.* Unicorn Stencil.

Gregory, C.A. & Coleman-Jensen, A. (2017, July). *Food insecurity, chronic disease, and health among working-Age adults.* U.S. Department of Agriculture, Economic Research Service.

Hall, J., Galarraga, J., Berman, I., Edwards, C., Khanjar, N., Kavi, L., Murray, R., Burwell-Naney, K., Jiang, C., & Wilson, S. (2021). Environmental injustice and industrial chicken farming in Maryland. *International Journal Environmental Research and Public Health, 18*(11039), 1–15. https://doi.org/10.3390/ijerph182111039

Hardoon, D. (2017). *An economy for the 99%. It's time to build a human economy that benefits everyone, not just the privileged few.* Oxfam International Briefing Paper. https://s3.amazonaws.com/oxfam-us/www/static/media/files/bp-economy-for-99-percent-160117-en.pdf

Health Equity Institute. (n.d.). *What is health equity.* San Francisco State University. http://healthequity.sfsu.edu

Ihejirika, M. (2023, May 24). *What is environmental racism?* National Resources Defense Council. www.nrdc.org/stories/what-environmental-racism

International Labour Organization, Walk Free & International Organization for Migration. (2022, September). *Global estimates of modern slavery: Forced labour and forced marriage.* www.ilo.org/sites/default/files/wcmsp5/groups/public/%40ed_norm/%40ipec/documents/publication/wcms_854795.pdf

James, W., Jia, C., & Kedia, S. (2012). Uneven magnitude of disparities in cancer risks from air toxics. *International Journal of Environmental Research and Public Health*, *9*, 4365–4385. https://doi.org/10.3390/ijerph9124365

Javidan, P. (2021). Structural sexism: Foundations, reproduction, and responses. In Leal Filho, W. et al. (Eds.), *Reduced inequalities, encyclopedia of the UN sustainable development goals* (pp 1–14). Springer Nature Switzerland AG. https://doi.org/10.1007/978-3-319-71060-0_90-1

Kaiser Family Foundation. (n.d.). *How history has shaped racial and ethnic health disparities. A timeline of policies and events.* KFF. www.kff.org/how-history-has-shaped-racial-and-ethnic-health-disparities-a timeline-of-policies-and-events/?entry=1808-to-1890-medical-exploitation-of-enslaved black-women

Kozol, J. (1992). *Savage inequalities*. HarperPerennial.

McCartney, G., Popham, F., McMaster, R., & Cumbers, A. (2019). Defining health and health inequalities. *Public Health*, *172*, 22–30. https://doi.org/10.1016/j.puhe.2019.03.023

Messenger, N. (2024a). *Giving away the future. How Louisiana's Industrial Tax Exemption Program (ITEP) has failed to produce prosperity.* Ohio River Valley Institute. https://ohiorivervalleyinstitute.org/wpcontent/uploads/2024/02/GivingAwaytheFuture_Final.pdf

Messenger, N., Hipple, K., & Keller, A. (2024b). *Pennsylvania's bad bet. Why Shell didn't save Appalachia with plastics.* Ohio River Valley Institute. https://ohiorivervalleyinstitute.org/wp-content/uploads/2024/01/Pennsylvanias-Bad-Bet-FINAL.pdf

Minnesota Pollution Control Agency. (2025). *East metro | 3M PFAS contamination.* www.pca.state.mn.us/local-sites-and-projects/east-metro-3m-pfas-contamination

National Association for the Advancement of Colored People. (2025). Criminal justice fact sheet. https://naacp.org/resources/criminal-justice-fact-sheet

National Oceanic and Atmospheric Administration. (2024a). *Coastal wetland habitat.* Office of Habitat Conservation. www.fisheries.noaa.gov/national/habitat-conservation/coastal-wetland-habitat

National Oceanic and Atmospheric Administration. (2024b). *$4.3 Million will support habitat restoration in coastal Louisiana.* Office of Habitat Conservation. www.fisheries.noaa.gov/feature-story/43-million-will-support-habitat-restoration-coastal-louisiana

Navas, G., DiAlisa, G., & Martinez-Alier, J. (2022). The role of working-class communities and the slow violence of toxic pollution in environmental health conflicts: A global perspective. *Global Environmental Change*, *73*(102474), 1–13. https://doi.org/10.1016/j.gloenvcha.2022.102474

Ndugga, N., & Artiga, S. (2023). *Disparities in health and health care: 5 key questions and answers.* Kaiser Family Foundation. www.kff.org/racial-equity-and-health-policy/issue-brief/disparities-in-health-and-health-care-5-key-question-and-answers/#

Nyéléni 2007 International Steering Committee. (2007). *Nyéléni 2007 Forum for Food Sovereignty.* Mali. https://nyeleni.org/DOWNLOADS/Nyelni_EN.pdf

O'Donnell, M. (2024, January 25). *Environmental justice and environmental racism.* https://instr.iastate.libguides.com/enviro_jr/ag

Oxfam America. (2015). *Lives on the line: The human cost of cheap chicken.* Oxfam America, Inc. https://s3.amazonaws.com/oxfamus/www/static/media/files/Lives_on_the_Line_Full_Report_Final.pdf

Prison Policy Institute. (n.d.). *Native incarceration in the U.S.* www.prisonpolicy.org/profiles/native.html#:~:text=Native%20people%20are%20incarcerated%20in,people%20(181%20per%20100%2C000).

Rabbitt, M.P., Hales, L.J., Burke, M.P., & Coleman-Jensen, A. (2023). *Household food security in the United States in 2022 (Report No. ERR-325).* U.S. Department of Agriculture, Economic Research Service. https://ers.usda.gov/sites/default/files/_laserfiche/publications/107703/ERR-325.pdf?v=66527

Sanchez, M., & Jameel, M. (2024, February 27). *Immigrant dairy workers often endure substandard housing conditions. The law doesn't protect them.* ProPublica. www.propublica.org/article/the-law-doesnt-protect-immigrant-dairy-workers-substandard-housing

Sharif, M.Z., Garcia, J.J., Mitchell, U., Dellor, E.D., Bradford, N.J., & Truong, M. (2022). Racism and structural violence: Interconnected threats to health equity. *Frontiers in Public Health, 9*(676783), 1–5. https://doi.org/10.3389/fpubh.2021.676783

Son, J.I., & Bell, M.L. (2021). Distribution of environmental justice metrics for exposure to CAFOs in North Carolina, USA. *Environmental Research, 195*(110862), 1–19. https://doi.org/10.1016/j.envres.2021.110862

Sunding, D.L. (2017, September 17). *Damage to Minnesota's natural resources resulting in resulting in 3M's disposal of PFC's in Washington County, MN.* The Brattle Group. Exhibit 89. Fourth Judicial Court.

Terrell, K.A., & St. Julien, G. (2022). Air pollution is linked to higher cancer rates among black or impoverished communities in Louisiana. *Environmental Research Letters, 17*(014033), 1–15. https://doi.org/10.1088/1748-9326/ac4360

The George Washington University. (2021). *Equity vs. equality? What's the difference?* Milken Institute School of Public Health. https://onlinepublichealth.gwu.edu/resources/equity-vs-equality/

Thomas, J.J. (2023). Equity, equality, and justice: Defined. *Seattle Journal for Social Justice, 21*(3), 713–715. https://digitalcommons.law.seattleu.edu/sjsj/vol21/iss3/13

United Nations. (n.d.). *Global issues: Water.* www.un.org/en/global-issues/water

United Nations Children's Fund (UNICEF), World Health Organization (WHO), International Bank for Reconstruction and Development/The World Bank. (2023). *Levels and trends in child malnutrition: UNICEF / WHO / World Bank Group joint child malnutrition Estimates: Key findings of the 2023 edition.* UNICEF & WHO. https://iris.who.int/bitstream/handle/10665/368038/9789240073791-eng.pdf?sequence=1

United States Department Apartment of Agriculture. (2009). *Access to affordable and nutritious food: Measuring and understanding food deserts and their consequences.* Economic Research Service. https://ers.usda.gov/sites/default/files/_laserfiche/publications/42711/12716_ap036_1_.pdf?v=70922

United States Department of Health and Human Services. (n.d.). *Social determinants of health.* https://health.gov/healthypeople/priority-areas/social-determinants-health

United States Department of Health and Human Services. (2022). *Health Equity and Health Disparities Environmental Scan.* Office of the Assistant Secretary for Health, Office of Disease Prevention and Health Promotion.

United States Environmental Protection Agency. (n.d.). *Animal feeding operations* (AFOs). National Pollutant Discharge Elimination System (NPDES). www.epa.gov/npdes/animal-feeding-operations-afos

United States Environmental Protection Agency. (2016, August 3). *Brownfield overview and definition.* US EPA. https://19january2017snapshot.epa.gov/brownfields/brownfield-overview-and-definition_.html

United States Environmental Protection Agency. (2023a). *What is Superfund?* US EPA. www.epa.gov/superfund/what-superfund

U.S. Environmental Protection Agency. (2023b). *Supporting environmental justice.* US EPA. www.epa.gov/superfund/supporting-environmental-justice-superfund-sites

United States Environmental Protection Agency. (2024). *About Brownfields.* UA EPA. www.epa.gov/brownfields/about

United States Environmental Protection Agency & Office of Land and Emergency Management. (2021, July). *Population surrounding 30,675 Brownfields sites.* www.epa.gov/system/files/documents/2021-10/brownfields.pdf

University Network for Human Rights. (2019, July). *"Waiting to die": Toxic emissions and disease near the Louisiana Denka/DuPont Plant.* www.humanrightsnetwork.org/waiting-to-die.

Widra, E., & Herring, T. (2021, September). *States of incarceration: The global context 2021.* Prison Policy Initiative. www.prisonpolicy.org/global/2021.html

Winter, D. (2022, December 15). *Toxic: 3M knew its chemicals were harmful decades ago, but didn't tell the public, government.* Minnesota Reformer. https://minnesotareformer.com/2022/12/15/toxic-3m-knew-its-chemicals-were-harmful-decades-ago-but-didnt-tell-the-public-government/

World Health Organization [WHO]. (n.d.). *Social determinants of health.* www.who.int/health-topics/social-determinants-of-health#tab=tab_1

Yavrom. (2021, April 28). An overview of hazardous waste generation. U.S. EPA. https://rcrapublic.epa.gov/rcra-public-web/action/posts/2

7 The Burden of Health Disparities and Nutrition-Related Chronic Disease in the U.S.

> "The lack of access to affordable nutritious food is associated with the increased risk for multiple chronic diseases; everyone in our nation needs access to healthy foods. Good nutrition makes for a stronger, healthier America."
>
> John Westerdahl[1], PhD, MPH, RDN, FAND, DipACLM

OVERVIEW

Neither pre-diabetes nor metabolic syndrome were recognized as health conditions when starting my career path as a clinical dietitian in 1993—but they are now recognized as serious health conditions that increase the risk of chronic diseases that are the leading causes of death globally. Similar to the global burden of chronic disease, nutrition-related chronic disease plagues the U.S., too. Furthermore, health disparities remain deeply concerning in the U.S., including the lack of access to health insurance and healthcare where

DOI: 10.1201/9781003153528-8

discrimination increases stressors and negative health outcomes. Geronimus et al. (2006) use the term "weathering" to describe the negative physiological outcomes that Black Americans face because of inequities, disparities, and sociocultural stressors. Hence, this chapter will focus on the interconnected issues associated with healthcare spending and the lack of access to healthcare; nutrition-related chronic disease and health disparities; the role of the built environment in health outcomes—and the relationship between COVID-19 outcomes and whole plant foods nutrition.

HEALTHCARE SPENDING IN THE U.S. AND ACCESS TO HEALTHCARE

The United States (U.S.) ranks far below wealthy and less wealthy countries in years of healthy life expectancy at birth (HALE) as denoted in Figure 7.1 despite spending the most on healthcare than any other country in the world (Rakshit et al., 2024; Wager et al., 2024). Comparing HALE solely to other wealthy nations, the U.S. has the lowest rank (Rakshit et al., 2024). In 2022, the life expectancy for a man in the U.S. was 74.8 years whereas for men in comparable nations, the average was 80 years; for a woman in the U.S., it was 80 years while women in comparable countries averaged over 84 years, respectively (Rakshit et al., 2024). Globally, Japan has the highest HALE that's attributed to decreased rates of cardiovascular disease, obesity, and cancer along with lower dietary intakes of red meat and sugar and higher intakes of soybeans, fermented soy products, green tea—and a diverse array of produce (Tsugane, 2021).

Although the Affordable Care Act decreased the number of Americans without health insurance in recent years alongside increased health provisions during the COVID-19 pandemic, challenges still remain for working families earning lower wages which may be compounded by where one lives (U.S. Department of Health and Health Services [US DHHS], 2023). As an example, Massachusetts had an uninsured rate of 2.5% in 2021 while Texas had an uninsured rate of 18% (US DHHS, 2023). The nearly 26 million people that were uninsured in the U.S. in 2022 were disproportionately people of color in working families with incomes that fell below the federal poverty line (Tolbert et al., 2023). With the exception of persons of Asian descent, Hispanics and American Indian/Alaskan Natives had an uninsured rate of over twice that of White Americans; Black Americans also experienced uninsured rates higher than whites (Tolbert et al., 2023). Those without health insurance have a decreased likelihood of accessing healthcare services (Tolbert et al., 2023).

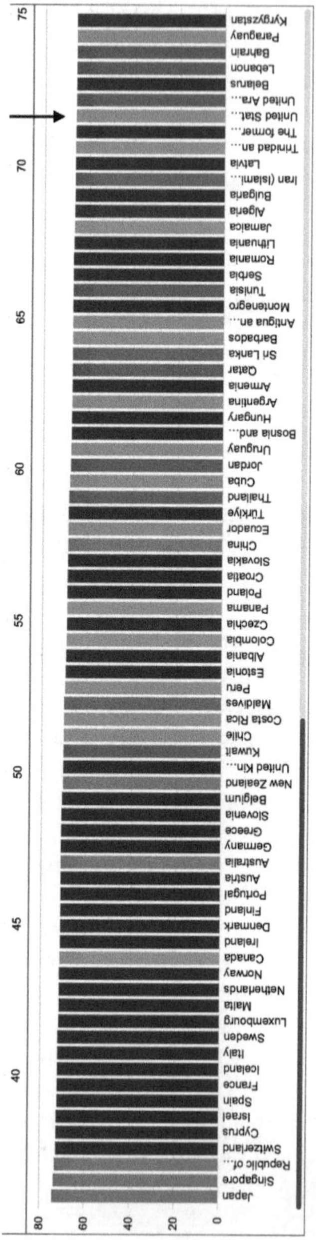

FIGURE 7.1 World Health Organization Global Health Observatory Healthy Life Expectancy at Birth (Years) With Arrow Demarcating the United States.

Furthermore, 43-51% of working adults with insurance coverage stated it was somewhat or very difficult to afford their healthcare—whether it was offered through their employer, a market plan, Medicaid, or Medicare while many were also encumbered with healthcare debts (Collins et al., 2023).

The high cost of healthcare in the U.S. compared to similar countries has not been attributed to the utilization of healthcare as Americans seek the services of a physician less than those in comparable countries (Gunja et al., 2023). The high cost of healthcare in the U.S. has been attributed in part to high administrative and healthcare billing costs as well as the high cost of prescription drugs that may be twice as much as comparable countries (Turner et al., 2023; Kurani & Cox, 2020). Shrank et al. (2019) assessed the literature between 2012–2019 regarding wasted healthcare monies that included failure of care delivery, failure of care coordination, overtreatment, pricing failure, fraud and abuse, and administrative complexity. Administrative complexity contributed the most to waste while approximately 25% of overall total healthcare spending is wasted—an estimated $760-935 billion dollars (Shrank et al., 2019). Healthcare administration costs per person in the U.S. is four times that of comparable nations (Kurani & Cox, 2020).

The U.S. is the only high-income country in the world that doesn't offer guaranteed health insurance where Americans' rate of obesity is twice that of persons living in Organization for Cooperation and Development (OECD) countries; Americans also live with the highest rates of multiple chronic conditions (Gunja et al., 2023).

CHRONIC DISEASES IN THE U.S.

The 2022 provisional statement on mortality in the U.S. released by the Centers for Disease Control and Prevention (CDC) identifies the top ten causes of death as: heart disease, cancer, unintentional injuries (largely attributed to drug-related overdoses), COVID-19, stroke, chronic lower respiratory diseases, Alzheimer's disease, diabetes, kidney disease, and chronic liver disease and cirrhosis (Ahmad et al., 2023). Mortality due to heart disease and cancer increased from the previous year while life expectancy decreased by 0.6 years for both men and women (Ahmad et al., 2023). Previously, researchers have calculated that future generations of Americans may have shortened lifespans due to the medical ramifications associated with obesity as well as other threats such as antibiotic resistance and infectious disease pandemics (Olshansky et al., 2005).

As it is, sixty percent (60%) of Americans live with one or more chronic diseases such as obesity, diabetes, cancer, stroke, and heart disease while 42% have two or more and 12% have at least five (Benavidez et al., 2024). Using tabulations for over 30,000 zip codes nationwide for the 10 most common and costly chronic diseases (obesity, hypertension, hypercholesterolemia, coronary artery disease, chronic obstructive pulmonary disease, diabetes, cancer, depression, chronic kidney disease, and stroke), Benavidez et al. (2024) found the highest prevalence of chronic disease was significantly associated with economic disadvantages such as high uninsured rates and barriers to health insurance, lower household income, lower educational attainment, and lower property values. Those with the highest prevalence of chronic disease were significantly more likely to be Alaska Native, American Indian, or Black (Benavidez et al., 2024).

Newman et al. (2020) assessed where Americans over 18 years of age who were diagnosed with two or more chronic diseases lived. The five states/territories having the highest prevalence included West Virginia (64.4%), Arkansas (60.5%), Alabama (60.1%), Kentucky (58.6%), and Maine (57.4%) (Newman et al., 2020). The states/territories with the lowest prevalence were the District of Columbia (37.9%), Minnesota (42.8%), Colorado (42.9%), California (43.5%), and Utah (43.7%), respectively (Newman et al., 2020).

Cardiovascular Disease and Stroke

Someone has a heart attack every 40 seconds and someone dies from cardiovascular disease every 33 seconds in the U.S. (Centers for Disease Control and Prevention [CDC], 2023a). Nearly half of all Americans have at least one risk factor for heart disease such as hypertension, high cholesterol levels, diabetes, or obesity—while sustained healthy lifestyle behaviors and whole plant foods nutrition can prevent, halt, or reverse heart disease (Ahluwalia et al., 2016; Esselstyn et al., 2014; Ornish et al., 1998). For Black Americans, Hispanic Americans, Alaska Native, and other Indigenous Americans, health disparities are evident; they are more likely to be diagnosed with obesity, coronary artery disease, and hypertension than White Americans, respectively (Lewsey & Breathett, 2021; Office of Minority Health, n.d.). A cohort study of over 220,000 Alaska Natives and American Indian Medicare beneficiaries found a substantial burden of cardiovascular disease and cardiometabolic risk factors where half were diagnosed with both severe cardiovascular disease and diabetes (Eberly et al., 2023).

Someone in the U.S. has a stroke every 40 seconds and just over every 3 minutes, someone dies of a stroke (CDC, 2023b). Stroke is the leading cause of long-term disability in the U.S. and Black persons experience their first stroke at twice the rate of White persons (CDC, 2023). Both Black persons and those of Pacific Island descent have the highest rates of mortality caused by stroke (CDC, 2023b).

Cancer

For the first time, it was projected in 2024 that the number of new cases of cancer in the U.S. will increase to over 2 million where lung, colorectal, and pancreatic cancers will claim the most lives (Siegel et al., 2024). Increasingly concerning is the rise in colorectal cancer cases in people less than 55 years of age (Siegel et al., 2024). Colorectal cancer is the leading cause of death amongst men 50 years of age or less where it's the second leading cause of death for women less than 50 years of age (Siegel et al., 2024). As previously discussed in Chapter 5, diet plays a critically important role in colorectal cancer.

For both women and men in the U.S., lung cancer is projected as the leading cause of mortality in 2024 (Siegel et al., 2024). The estimated leading cause of new cancer cases for women is breast cancer and for men it's prostate cancer, respectively (Siegel et al., 2024). For both genders, the second leading cause of new cancer cases is lung cancer followed by colorectal cancer (Siegel et al., 2024). Health disparities persist as well where the risk of cancer mortality is 51% higher for American Indian and Alaska Natives and 33% higher for Black individuals than White individuals, respectively (Siegel et al., 2024). In the case of endometrial cancer, mortality amongst Black women is double that of White women and for breast cancer, Black women experience a 41% higher mortality rate (Siegel et al., 2024).

Hastert and White (2016) assessed whether six lifestyle recommendations set forth by the World Cancer Research Fund (WCRF) & the American Institute for Cancer Research (AICR) impacted colorectal cancer outcomes in a cohort study of over 66,000 adults aged 50-76. The recommendations assessed included: eating greater than or equal to 5 servings of a wide variety of non-starchy fruits and vegetables a day; eating a serving of unprocessed grains and legumes at each meal; limiting alcohol intake to one drink a day for women and no more than two drinks per day for men; achieving a body mass index between 18.5-25 kg/m^2; engaging in a brisk 30-minute walk daily 5 days/week over the past 7-10 years; limiting energy-dense foods such as added sugars and fast foods; and limiting red and processed meats to 18 ounces/week (Hastert & White, 2016). Overall, meeting 4 to 6 of these recommendations lowered the risk of colorectal cancer by 58% whereas only meeting 1 to 3 of the recommendations lowered the risk by 34% to 45%, respectively (Hastert & White, 2016). Additionally, following the alcohol recommendations was associated with a 19% lower colorectal cancer risk while meeting Body Mass Index (BMI) guidelines decreased the risk by 15%, respectively (Hastert & White, 2016).

The 2018 WCRF & AICR report entitled *Diet, Nutrition, Physical Activity and Colorectal Cancer* demonstrated that physical activity, whole grains, high fiber foods, and adequate calcium intake convincingly decreased the risk of colorectal cancer whereas processed meats, red meat, obesity, and drinking greater than two alcoholic drinks per day convincingly increase the risk of colorectal cancer (World Cancer Research Fund [WCRF] & American Institute for Cancer Research [AICR], 2018). Two-thirds of colorectal cancers occur in high income nations where the Western diet and sedentary lifestyle prevail; only 5-10% of colorectal cancer are thought to be due to a "recognized hereditary condition"; 20% are associated with a family history (WCRF & AICR, 2018). Hence, the majority of colorectal cancer cases are not inevitable, but preventable.

Hypertension, Obesity, and Metabolic Syndrome

Almost half of all Americans have hypertension known as the "silent killer" which increases the risk of heart disease, stroke, kidney disease, vision loss, and other vascular problems (Fryar et al., 2024; American Heart Association, 2022). Only 25% of Americans with hypertension are effectively controlling it (CDC, 2023c). Analyzing National Health and Nutrition Examination Survey (NHANES) data from over 55,000 Americans, the cardiometabolic health of Americans is declining where only ~7% of Americans experience optimal cardiometabolic health; health disparities were evident among persons of lower income, Black adults, and Hispanic adults (O'Hearn et al., 2022). Sells et al. (2023) analyzed NHANES data from 1999–2018 with over 31,000 subjects aged 30 or older and found that Black Americans had the highest prevalence of hypertension compared to white and Hispanic subjects, respectively. When assessing the impact of poverty on hypertension and cardiovascular disease, disease burden was exacerbated amongst Blacks and Hispanics (Sells et al., 2023). While it's been known for decades now that the DASH diet (Dietary Approaches to Stop Hypertension) containing whole grains, fruits, vegetables, etc., along with healthy lifestyle behaviors can decrease blood pressure—there are systemic failures in disease management (Filippou et al., 2020; O'Hearn et al., 2022).

Nearly 73% of American adults are overweight or obese while the prevalence of severe obesity is over 9% (National Institutes of Diabetes and Digestive and Kidney Diseases [NIDDK], 2021). Black Americans have the highest rates of obesity which is followed by Hispanic adults as seen in Figure 7.2 (Petersen et al., 2019). Between 1999–2020, Black Americans also experienced the highest age-adjusted mortality from obesity-related cardiovascular disease than Asian or Pacific Islanders, American Indian or Alaska Natives, and White persons, respectively (Raisi-Estabragh et al., 2023). The Office of Minority

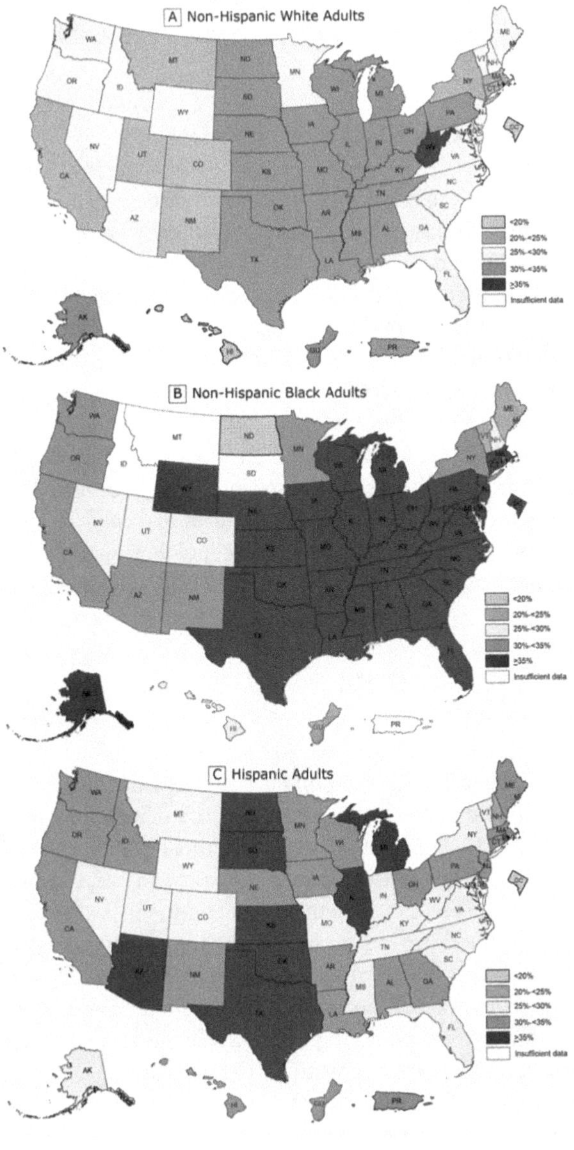

FIGURE 7.2 The Prevalence of Self-Reported Obesity by U.S. State and Territory Amongst A) White Adults (non-Hispanic) | B) Black Adults (non-Hispanic) | C) Hispanic Adults, 2017 Behavioral Risk Factor Surveillance System. Peterson et al., 2019, Centers for Disease Control and Prevention.

Health reports that Native Americans and Alaska Natives are 40% more likely to be obese than White Americans (Office of Minority Health, 2024).

Type 2 Diabetes and Pre-Diabetes

Diabetes can be a seriously debilitating chronic disease and is a risk factor for heart disease, renal disease, renal failure, neuropathy, stroke, loss of limbs or sight, etc. (NIDDK, n.d.). Boyle et al. (2010) projected that up to one-third of Americans could be living with diabetes by 2050 where another analysis by Lin et al. (2018) found that the number of adults diagnosed with diabetes could triple between the years 2014–2060. One in 5 Americans has diabetes but doesn't know it (CDC, 2022). In 2021, more than 1 in 3 persons in the U.S. had pre-diabetes (NIDDK, 2024).

Between 2019–2021, American Indian and Alaska Natives, Asian, Black, and Hispanic Americans were diagnosed with diabetes at higher rates than White Americans (CDC, 2022). Education level and income also play a role where higher income and higher levels of education were associated with a lower rate of diabetes diagnosed between 2019–2021 (CDC, 2022).

The Built Environment

The U.S. Environmental Protection Agency (EPA) defines the built environment as, "The man-made or modified structures that provide people with living, working, and recreational spaces" (U.S. Environmental Protection Agency [EPA], 2024). The built environment impacts the Social Determinants of Health (SDH) and health outcomes in a variety of ways including but not limited to the lack of access to green spaces and the lack of trees which offer shade; indoor or outdoor toxicants; substandard housing; increased urban heat levels; psychological or physical violence; the inability to engage in physical activity or active living outdoors; obesogenic food environments; the lack of reliable public transportation; polluting businesses; substandard utility services; and neighborhood segregation (Nabaweesi et al., 2023; National Academy of Sciences, 2023).

Urban heat islands are areas that are generally (but not exclusively) metropolitan that have higher temperatures due to the heat-absorbing properties of impervious surfaces (Climate Central, 2021). They have fewer trees, less vegetation, and taller buildings (Climate Central, 2021). The impact of urban heat islands is exacerbated by climate change and record-setting heat (Climate Central, 2021). Heat-related deaths may be due to the extreme heat itself or a contributing factor such as a heat-related cardiovascular event (US EPA, 2023).

Those that are at higher risk during extreme heat include individuals that: have pre-existing health conditions; are older adults; live alone; have limited access to or do not have healthcare resources; are of lower socioeconomic status; use diuretics and other medicines that either dehydrate or dysregulate body cooling functions; reside in urban heat islands; and work outdoors (US DHHS, 2024). Agricultural workers are particularly vulnerable due to a lack of safety regulations, extremely low wages and wage structures that incentivize productivity over safety, language barriers, immigration status, etc. (Environmental Defense Fund, 2023).

Using the neighborhood appraisal maps that led to redlining in the U.S., Locke et al. (2021) found a significant difference between tree canopy and neighborhood appraisal status; areas with higher neighborhood appraisal scores had twice the tree cover than communities of color which had lower neighborhood appraisal scores (Locke et al., 2021). Furthermore, significant urban heat disparities have been noted in the U.S. where low-income communities and communities of color experience higher land-surface temperatures compared to wealthier neighborhoods (Benz & Burney, 2021). Racism has negatively shaped urban environments in the U.S. where communities of color and low-income communities along with the natural environment experience increased burdens (Benz & Burney, 2021; Nabaweesi et al., 2023).

COVID-19, HEALTH DISPARITIES, AND DIET

The COVID-19 pandemic demonstrated that health disparities, race, ethnicity, and multiple chronic health conditions were associated with severe complications or mortality (CDC, 2023d). Death from COVID-19 disproportionately impacted people of color along with impoverished communities (CDC, 2023d). Importantly, diet made a difference in COVID-19 severity and outcomes. Researchers in Brazil found that across ~700 study participants, those consuming largely vegetarian dietary patterns that included higher intakes of vegetables, legumes, and nuts had a lower incidence of COVID-19 (Acosta-Navarro et al., 2024). Kim et al. (2021) assessed the diets of healthcare workers with high exposure to COVID-19 in six countries and found that participants consuming more whole-plant foods decreased the odds of COVID-19 severity. Finally, Merino et al. (2021) surveyed persons living in the United Kingdom and the U.S. and found that intake of healthy plant foods was associated with decreased risk of and decreased severity related to COVID-19. These findings speak to the critically important role of whole-plant foods nutrition in immune function and anti-inflammatory processes as well as the importance of access to healthy plant foods for vulnerable individuals and communities (Khalid et al., 2022).

NOTE

1 Westerdahl, J. (2024, June 8). Personal communication.

REFERENCES

Acosta-Navarro J.C., Dias, L.F., de Gouveia, L.A.G., Ferriera, E.P., de Oliveira, N.V.P.F., Marin, F.A., Oiveira, J.V.C., da Silva, A.C., Silva, I.L., Freitas, F.D., & Soares, P.R. (2024). Vegetarian and plant-based diets associated with lower incidence of COVID-19. *BMJ Nutrition, Prevention & Health*, *7*(1), 4–13. https://doi.org/10.1136/bmjnph-2023-000629

Ahluwalia, Gladwin, M., Coleman, G.D., Hord, N., Howard, G., Kim-Shapiro, D.B., Lajous, M., Larsen, F.J., Lefer, D.J. McClure, L.A., Nolan, B.T., Pluta, R., Schrechter, A., Wang, C., Ward, M.H., & Harman, J.L. (2016). Dietary nitrate and the epidemiology of cardiovascular disease: Report from a National Heart, Lung, and Blood Institute workshop. *Journal of the American Heart Association*, *5*(7), 1–10. https://doi.org/10.1161/JAHA.116.003402

Ahmad, F.B., Cisweski, J.A., Xu, J., & Anderson, R.N. (2023). Provisional mortality data — United States, 2022. *Morbidity and Mortality Weekly Report*, *72*(18), 488–492.

American Heart Association. (2022, March 4). *Health threats from high blood pressure*. www.heart.org/en/health-topics/high-blood-pressure/health-threats-from-high-blood-pressure

Benavidez, G.A., Zahnd, W.E., Hung, P., & Eberth, J.M. (2024). Chronic disease prevalence in the US: Sociodemographic and geographic variations by zip code tabulation area. *Preventing Chronic Disease*, *21*(e14), 1–10. http://doi.org/10.5888/pcd21.230267

Benz, S.A., & Burney, J.A. (2021). Widespread race and class disparities in surface urban heat extremes across the United States. *Earth's Future*, *9*(7), 1–14. https://doi.org/10.1029/2021EF002016

Boyle, J.P., Thompson, T.J., Gregg, E.W., Barker, L.E., & Williamson, D.F. (2010). Projection of the year 2050 burden of diabetes in the US adult population: Dynamic modeling of incidence, mortality, and prediabetes prevalence. *Population Health Metrics*, *8*(29), 1–12.

Centers for Disease Control and Prevention. (2022, October 25). *By the numbers: Diabetes in America.* www.cdc.gov/diabetes/health-equity/diabetes-by-the-numbers.html

Centers for Disease Control and Prevention. (2023a). *Heart disease facts.* www.cdc.gov/heartdisease/facts.htm

Centers for Disease Control and Prevention. (2023b). Stroke facts. www.cdc.gov/stroke/facts.htm

Centers for Disease Control and Prevention. (2023c). *Estimated hypertension prevalence, treatment, and control among U.S. adults.* https://millionhearts.hhs.gov/data-reports/hypertension-prevalence.html

Centers for Disease Control and Prevention. (2023d). *Underlying medical conditions associated with higher risk for severe COVID-19: Information for healthcare professionals.* www.cdc.gov/coronavirus/2019-ncov/hcp/clinical-care/underlyingconditions.html

Climate Central. (2021, July 14). *Hot zones: Urban heat islands.* https://assets.ctfassets.net/cxgxgstp8r5d/1XZZjkLYwtcmKL5k3wEinl/5f8c9b5b2d8dd56e1bda7f51278fc3d2/2021_UHI_Report.pdf

Collins, S.R., Roy, S., & Masitha, R. (2023, October). *Paying for it: How health care costs and medical debt are making Americans sicker and poorer —findings from the Commonwealth Fund 202 Health Care Affordability Survey.* Commonwealth Fund. https://doi.org/10.26099/bf08-3735

Eberly, L.A., Shultz, K., Merino, M., Breuckner, M.Y., Benally, E., Tennison, A., Biggs, S., Hardie, L., Tian, Y., Nathan, A.S., Khatana, S.A.M., Shea, J.A., Lewis, E. Bukham, G., Shin, S., & Groeneveld, P.W. (2023). Cardiovascular disease burden and outcomes among American Indian and Alaska native Medicare beneficiaries. *JAMA Network Open*, *6*(9), e2334923, 1–13. https://doi.org/10.1001/jamanetworkopen.2023.34923

Environmental Defense Fund. (2023). *Farmworkers and Heat Stress in the United States: A future proofing in U.S. agriculture report.* www.edf.org/sites/default/files/2023-07/EDF040_Heat%20Stress_V5.pdf

Esselstyn, C.B, Gendy, G., Doyle, J., Golubic, M., & Roizen, M.F. (2014). A way to reverse CAD? *Journal of Family Practice*, *63*(7), 356–364.

Filippou, C.D., Tsioufis,C.P., Thomopoulos, C.G., Mihas, C.C., Dimitriadis, K.S., Sotiropoulou, L.I., Chrysochoou, C., Nihoyannopoulos, P.I., & Tousoulis, D.M. (2020). Dietary Approaches to Stop Hypertension (DASH) diet and blood pressure reduction in adults with and without hypertension: A systematic review and meta-analysis of randomized controlled trials. *Advances in Nutrition*, *11*, 1150–1160. https://doi.org/10.1093/advances/nmaa041

Fryar, C.D., Kit, B., Carroll, M.D., & Afful, J. (2024). *Hypertension prevalence, awareness, treatment, and control among adults age 18 and older: United States, August 2021– ugust 2023. Data Brief No 511.* National Center for Health Statistics. https://dx.doi.org/10.15620/cdc/164016

Geronimus, A.T., Hicken, M., Keene, D., & Bound, J. (2006). "Weathering" and age patterns of allostatic load scores among blacks and whites in the United States. *American Journal Public Health*, *96*(5), 826–833. https://dx.doi.org/10.2105/AJPH.2004.060749

Gunja, M.Z., Gumas, E.D., & Williams, R.D. (2023, January 21). *U.S. health care from a global perspective, 2022: Accelerating spending, worsening outcomes.*

Commonwealth Fund. www.commonwealthfund.org/publications/issue-briefs/2023/jan/us-health-care-global-perspective-2022

Hastert, T.A., & White, E. (2016). Association between meeting the WCRF/AICR cancer prevention recommendations and colorectal cancer incidence: Results from the VITAL cohort. *Cancer Causes & Control*, *27*(11), 1347–1359. https://dx.doi.org/10.1007/s10552-016-0814-6

Khalid, W., Arshad, M.S., Ranjha, M.M., Różańska, M.B., Irfan, S., Shafique, B., Rahim, M.A., Khalid, M.Z., Abdi, G., & Kowalczewski, P.L. (2022). Functional constituents of plant-based foods boost immunity against acute and chronic disorders. *Open Life Sciences*, *17*, 1075–1093. https://doi.org/10.1515/biol-2022-0104

Kim, H., Rebholz, C.M., Hedge, S., LaFiura, C., Raghavan, M., Floyd, J.F., Cheng, S., & Seidelmann, S.B. (2021). Plant-based diets, pescatarian diets and COVID-19 severity: A population-based case–control study in six countries. *BMJ Nutrition, Prevention & Health*, *4*(1), 257–266. https://doi.org/10.1136/bmjnph-2021-000272

Kurani, N., & Cox, C. (2020, September 25). *What drives health spending in the U.S. compared to other countries.* Peterson-KFF Health System Tracker. www.healthsystemtracker.org/brief/what-drives-health-sp…e%20spending%20per%20capita,%20by%20spending%20category,%202018

Lewsey, S.C., & Breathett, K. (2021). Racial and ethnic disparities in heart failure: Current state and future directions. *Current Opinion in Cardiology*, *36*(3), 320–328. https://doi.org/10.1097/HCO.0000000000000855

Lin, J., Thompson, T.J., Cheng, Y.J., Zhuo, X., Zhang, P., Gregg, E., & Rolka, D.B. (2018). Projection of the future diabetes burden in the United States through 2060. *Population Health Metrics*, *16*(9), 1–9. https://doi.org/10.1186/s12963-018-0166-4

Locke, D., Hall, B., Grove, J.M. Pickett, S.T.A., Odgen, L.A., Aoki, C., Boone, C.G., & O'Neil-Dunne, J.P.M. (2021). Residential housing segregation and urban tree canopy in 37 US Cities. *Urban Sustainability*, *15*, 1–9.

Merino, N., Stock, W., Rousseau, C., Dierickx, G., D'hondt, S., Martínez, L. A., Dittami, S.M., van der Loos, L.M., & De Clerck, O. (2021). Breaking free from references: A consensus-based approach for community profiling with long amplicon nanopore data. *Briefings in Bioinformatics*, *22*(5), bbab120. https://doi.org/10.1093/bib/bbab120

Nabaweesi, R., Hanna, M., Muthuka, J.K., Samuels, A.D., Brown, V., Schwartz, D., & Ekadi, G. (2023). The built health environment as a determinant of health. *Primary Care: Clinics in Office Practice*, *50*(4), 591–599. https://doi.org/10.1016/j.pop.2023.04.012

National Academies of Sciences, Engineering, and Medicine. (2023). *Federal policy to advance racial, ethnic, and tribal health equity*. The National Academies Press. https://doi.org/10.17226/26834.

National Institutes of Diabetes and Digestive and Kidney Diseases [NIDDK]. (n.d.). *Preventing diabetes problems.* www.niddk.nih.gov/health-information/diabetes/overview/preventing-problems

National Institutes of Diabetes and Digestive and Kidney Diseases [NIDDK]. (2021, September). *Overweight and obesity statistics.* www.niddk.nih.gov/health-information/health-statistics/overweight-obesity

National Institutes of Diabetes and Digestive and Kidney Diseases [NIDDK]. (2024, January). *Diabetes statistics.* www.niddk.nih.gov/health-information/health-statistics/diabetes-statistics

Newman D., Tong, M., Levine, E., & Kishore, S. (2020). Prevalence of multiple chronic conditions by U.S. state and territory, 2017. *PLoS ONE, 15*(5), 1–12. https://doi.org/10.1371/journal.pone.023234

Office of Minority Health. (n.d.). *Heart disease and American Indians/Alaska Natives.* US DHHS. https://minorityhealth.hhs.gov/heart-disease-and-american-indiansalaska-natives

Office of Minority Health. (2024, October 29). *Obesity and American Indians/Alaska Natives.* US Department of Health and Human Services [US DHHS]. https://minorityhealth.hhs.gov/obesity-and-american-indiansalaska-natives

O'Hearn, M., Lauren, B.N., Wong, J.B., Kim, D.D., & Mozaffarian, D. (2022). Trends and disparities in cardiometabolic health among U.S. adults, 1999–2018. *Journal of the American College of Cardiology, 80*(2), 138–151. https://doi.org/10.1016/j.jacc.2022.04.046

Olshansky, S., Passaro, D.J., Hershow, R.C., Layden, J., Carnes, B.A, Brody, J., Hayflick, L., Butler, R.N, Allison, D.B. & Ludwig, D.S. (2005). A Potential Decline in Life Expectancy in the United States in the 21st Century. *New England Journal of Medicine, 352*(11), 1138–1145.

Ornish, D., Scherwitz, L.W., Billings, J.H., Gould, L., Merritt, T.A., Sparler, S., Armstrong, W.T., Ports, T.A., Kirkeeide, R.L., Hogeboom, C., & Brand, R.J. (1998). Intensive lifestyle changes for reversal of coronary heart disease. *JAMA, 280*(23), 2001–2008.

Peterson, R., Pan, L., & Blanck, H.M. (2019). Racial and ethnic disparities in adult obesity in the United States: CDC's tracking to inform state and local action. *Preventing Chronic Disease, 16*(E46), 1–6. http://doi.org/10.5888/pcd16.180579

Raisi-Estabragh, Z., Kobo, O., Mieres, J.H., Bullock-Palmer, R.P., Van Spall, H.G.C., Breathett, K., & Mamas, M.A. (2023). Racial disparities in obesity-related cardiovascular mortality in the United States: Temporal trends from 1999 to 2020. *Journal of the American Heart Association, 12*, 1–13. https://doi.org/10.1161/JAHA.122.028409

Rakshit, S., McGough, M., & Amin, K. (2024, January 30). *How does U.S. life expectancy compare to other Countries?* Peterson-Kaiser Family Foundation Health Tracker. www.healthsystemtracker.org/chart-collection/u-s-life expe…ntries/#Life%20expectancy%20at%20birth,%20in%20years,%201980-2022

Sells, M.L., Blum, E., Perry, G.S., Eke, P., & Presley-Cantrell, L. (2023). Excess burden of poverty and hypertension, by race and ethnicity, on the prevalence of cardiovascular disease. *Preventing Chronic Disease, 20*(230065), 1–13. https://doi.org/10.5888/pcd20.230065

Siegel, R.L., Giaquinto, A.N., & Jemal, A. (2024). Cancer statistics, 2024. *CA: A Cancer Journal for Clinicians, 74*, 12–49. https://doi.org/10.3322/caac.21820

Shrank, W.H., Rogstad, T.L., & Parekh. (2019). Waste in the US health care system estimated costs and potential for savings. *JAMA*, *322*(15), 1501–1509. http://doi.org/10.1001/jama.2019.13978

Tolbert, J., Drake, P.D., & Damico, A. (2023, December 18). *Key facts about the uninsured population*. Kaiser Family Foundation. www.kff.org/uninsured/issue-brief/key-facts-about-the-uninsured-population/view/print/

Tsugane, S. (2021). Why has Japan become the world's most long-lived country: Insights from a food and nutrition perspective. *European Journal of Clinical Nutrition*, *75*, 921–928. https://doi.org/10.1038/s41430-020-0677-5

Turner, A. Miller, G., & Lowry, E. (2023, October 4). *High U.S. health care spending: Where is it all going?* Commonwealth Fund. https://doi.org/10.26099/r6j5-6e66

U.S. Department of Health and Human Services. (2024, April 3). *Extreme heat*. www.hhs.gov/climate-change-health-equity-environmental-justic…-change-health-equity/climate-health-outlook/extreme-heat/index.html

U.S. Department of Health and Health Services [US DHHS]. (2023, August). *National uninsured rate reaches an all-time low in early 2023 after close of ACA open enrollment period* (Issue Brief No. HP-2023-20). Office of the Assistant Secretary for Planning and Evaluation.

U.S. Environmental Protection Agency. (2024, February 8). *Basic Information about the built environment*. www.epa.gov/smm/basic-information-about-built-environment#builtenviron

U.S. Environmental Protection Agency. (2023, November 1). Climate change indicators: Heat-related deaths. www.epa.gov/climate-indicators/climate-change-indicators-heat-related-deaths

Wager, E., McGough, M., Rakshit, S., & Amin, K. (2024, January 23). *How does health spending in the U.S. compare to other countries?* Peterson-Kaiser Family Foundation Health Tracker. www.healthsystemtracker.org/chart-collection/health-spen…ding%20per%20capita,%202022%20(U.S.%20dollars,%20PPP%20adjusted)

World Cancer Research Fund & American Institute for Cancer Research. (2018). *Diet, nutrition, physical activity and cancer: A global perspective. Continuous Update Project expert report 2018*. www.dietandcancerreport.org

World Health Organization. (2024). *The global health observatory*. www.who.int/data/gho/data/indicators/indicator-details/GHO/gho-ghe-hale-healthy-life-expectancy-at-birth

SECTION TWO

Multifaceted Solutions Harnessing Nutrition, Nature and Equity

One Health for All 8

"By protecting one, we help protect all."
Food and Agriculture Organization of the United Nations,
One Health Day Slogan[1]

OVERVIEW

The second half of *Nutrition Security for Planetary Health* focuses on solutions to the multifaceted planetary health crises. This chapter starts by defining the One Health concept and introduces nature-based thinking and Indigenous wisdom. Thereafter, the importance of bolstering nature with credible nature-based solutions is reinforced where regenerative agroecology is vital. Integrated topics include dry farming, orphan crops, wild plants, the importance of native plant ecosystems—and examples of traditional/Indigenous agroecological practices that create nutrient security and bolster planetary health.

DEFINING THE ONE HEALTH APPROACH

The World Health Organization (WHO) et al. (2023) defines One Health as, "an integrated, unifying approach that aims to sustainably balance and optimize the health of people, animals and ecosystems. It recognizes the health of humans, domestic and wild animals, plants, and the wider environment (including ecosystems) are closely linked and interdependent. The approach mobilizes multiple sectors, disciplines and communities at varying levels of society to work together to foster well-being and tackle threats to health and ecosystems, while addressing the collective need for clean water, energy and air, safe and nutritious food, taking action on climate changes and contributing

DOI: 10.1201/9781003153528-10

to sustainable development" (World Health Organization [WHO] et al., 2023). The Centers for Disease Control and Prevention (CDC) defines One Health as, "...a collaborative, multisectoral, and transdisciplinary approach—working at the local, regional, national, and global levels—with the goal of achieving optimal health outcomes recognizing the interconnection between people, animals, plants, and their shared environment" (Centers for Disease Control and Prevention [CDC], 2023). Additionally, the American Public Health Association (APHA) recognizes the One Health approach as a cost-effective strategy in transforming the immense burden of disease and environmental degradation (American Public Health Association [APHA], 2017). Figure 8.1 demonstrates the all-encompassing nature of the One Health Concept.

Enacting the One Health concept is a dramatic departure from the prevailing practices of global policies and the industrialized food system. As the COVID-19 pandemic was yet another wake-up call, the only way to prevent future pandemics, keep antibiotics working, ameliorate chronic disease, achieve food security, create justice, and end damage to the Earth System is enacting the One Health concept. The underlying principles of the One Health concept includes equity, sociocultural and multicultural parity, socioecological equilibrium, stewardship, cross disciplinary collaboration, and capacity building at all levels to build healthy ecosystems, healthy humans, and healthy animals as depicted in Figure 8.2 (One Health High-Level Expert Panel [OHHLEP] et al., 2022).

Applying the One Health Approach in Industrial Food-Animal Production

An example of applying the One Health concept in practice in the U.S. comes from the burdens associated with the mainstream food-animal industry where multifaceted and integrated solutions are necessary. The APHA has called for a precautionary moratorium on new or expanding Concentrated Animal Feeding Operations (CAFOs) (APHA, 2019). The U.S. Environmental Protection Agency (EPA) defines CAFOs by the type of animal and the respective manure handling system where a small CAFO ranges from 300-37,500 animals; a large CAFO can consist of 500-125,000 animals (U.S. Environmental Protection Agency [EPA], n.d.).

The APHA's recommended actions includes: mandating compliance of antibiotic use that coincides with the WHO's guidance; upholding laws such as the Clean Air Act, the Clean Water Act, the Occupational Safety and Health Act, and Title VI of the 1964 Civil Rights Act; phasing out liquid manure handling systems and waste lagoons and prohibiting any others; implementing and enforcing a strict nutrient management plan; eliminating subsidies and

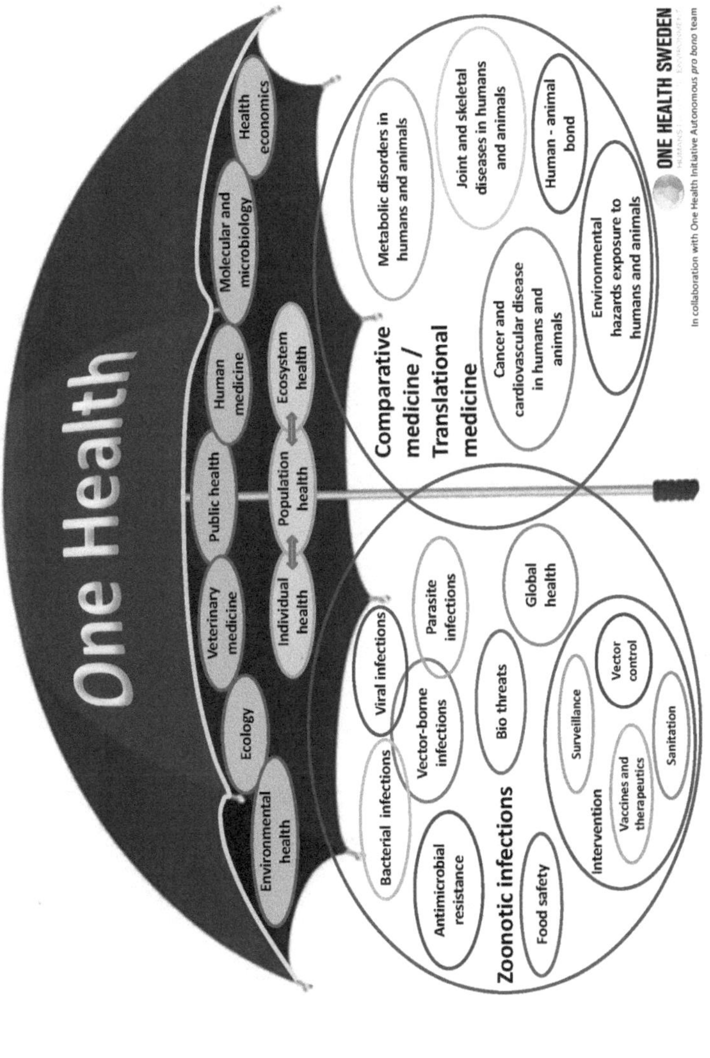

FIGURE 8.1 One Health Concept Designed by One Health Sweden & One Health Initiative Autonomcus Pro Bono Team.

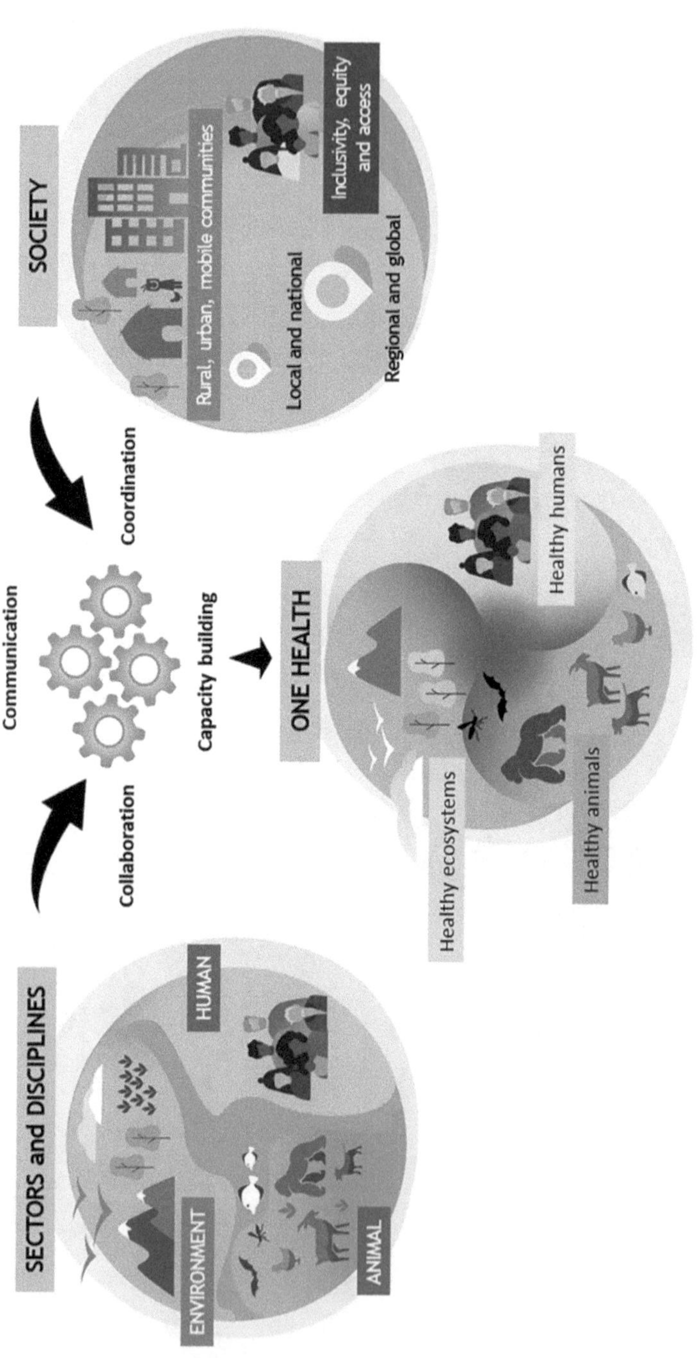

FIGURE 8.2 One Health High-Level Expert Panel "Towards a Sustainable Healthy Future", 2022.

reallocating monetary resources; and allowing zoning laws to protect communities (APHA, 2019). Nitzan et al. (2023) recommend expanding the One Health concept to *prioritize* food systems. Others such as U.S. Senator Cory Booker (New Jersey) and U.S. Representatives Dr. Alma Adams (North Carolina) and Jim McGovern (Massachusetts) introduced a proposed federal bill in September 2024 entitled, "The Industrial Agriculture Conversion Act" that would empower farmers to transition away from the CAFO model (Booker, 2024). All of these strategies employ actions to end business-as-usual practices, embracing One Health.

NATURE-BASED THINKING AND INDIGENOUS WISDOM

Mercado et al. (2024) offer the concept of "nature-based thinking" as a thoughtful approach that moves beyond a human-centered focus of nature. Instead, the focus is *nature with people* versus nature *for* people (Mercado et al., 2024). Nature-based thinking is a reflective process that fosters reciprocal, long-term relationships between humans and nature where both local and traditional knowledge is utilized (Mercado et al., 2024). In nature-based thinking, the worth and rights of nature are inherent and are not based on their respective value to humans (Mercado et al., 2024). Indigenous persons consider the "personhood of nature" an ingrained way of life—and a determinant of planetary health (Redvers et al., 2022). For example, rivers have the right to good health and flow as they were meant to.

A diverse array of Indigenous cultures believe they are inseparable from the Earth which defines the interconnectedness of all things (Reed et al., 2024). Canadian Indigenous leaders believe that the continued commodification of the natural world reinforces the destruction of the natural environment, the legacy of colonialism, and extractive capitalism (Reed et al., 2024). As global policies are devoid of humanity's reciprocal relationship with nature, incorporating Indigenous wisdom offers "life-enhancing" solutions (Reed et al., 2024).

Also important are the Indigenous values of collective versus individualistic societies where "Natural Laws" or "First Laws" take precedence. Natural Laws/First Laws create a cooperative society grounded in love and reciprocity where the rights of all ancestors including the First Mother (Mother Earth) are prioritized (Redvers et al., 2020). Importantly, the mainstream disconnect between humans and their indebtedness to the First Mother must be reversed to restore planetary health (Redvers et al., 2022).

NATURE-BASED SOLUTIONS

As discussed in Chapter 6, wetlands are recognized for their invaluable ecosystem services, including storm and flood protection. Hurricane Katrina is the deadliest storm on record in the U.S. (National Centers for Environmental Information, 2024). Such catastrophic climate events are likely thought of as "natural disasters" where the destruction is seemingly inevitable and beyond the control of humans (Smith, 2005). However, Hurricane Katrina revealed epic human errors that started well before the time of impact and included the systematic and accelerated degradation of the natural environment; engineering mistakes; funding deficiencies—and longstanding racism (Barbier, 2013; Center for Progressive Reform, 2005; Smith, 2005). Scientists have concluded that wetland restoration in southeast Louisiana is the best path forward to prevent future climate disasters (Barbier, 2013). Thus, restoring and rejuvenating ecosystems and biodiversity to their wild and natural processes is an essential solution and another example of One Health.

"Nature-based solutions" are being heavily touted as a solution to the multifaceted environmental crises. Nature-based solutions can be defined as "actions to protect, sustainably manage and restore natural and modified ecosystems in ways that address societal challenges effectively and adaptively, to provide both human well-being and biodiversity benefits" (International Union for the Conversation of Nature, 2020). Nature-based solutions work because if left unperturbed, nature's integrated living systems and biodiversity underpin life on this planet.

However, it depends on who is defining or devising the nature-based intervention (Melandis & Hagerman, 2022). An analysis of California's forest carbon offset program (a "cap-and-trade" program) found that it didn't improve climate mitigation outcomes (Badgley et al., 2021). Another in-depth analysis of voluntary carbon markets found widespread flaws and the need for reform (Battocletti et al., 2023). Furthermore, West et al. (2023) analyzed data from 26 carbon offset projects on three continents associated with mitigating deforestation and found that most projects did not significantly reduce deforestation.

REGENERATIVE AGROECOLOGY, ORGANIC FARMING, AND CONSERVATION FARMING

Applying One Health to the entirety of agriculture is vital. Conventional farming (mainstream agriculture) is characterized by the reliance on fossil fuels, machinery, and the intense manipulation of our living systems (Intergovernmental Science-Policy Platform on Biodiversity and Ecosystem Services [IPBES], n.d.). In contrast, "regenerative sustainability" describes holistically changing the dominant view of agriculture where living systems can thrive (Gibbons, 2020). Regenerative agriculture uses practices that nourish and replenish ecosystems and includes food sovereignty (Gibbons, 2020).

Agroecologists have long advocated for and have demonstrated that producing food in ways that mirror the natural processes of living systems strengthens ecosystem resilience (Altieri & Nicholls, 2020; Perfecto et al., 2009). This means not only refraining from agricultural processes that disrupt natural cycles—but nourishing the Earth's biology by creating and nurturing soil fertility; incorporating indigenous, local, and traditional wisdom into agricultural practices; phasing out agrichemicals and using integrated pest management; ensuring crop diversity/intercropping, etc. (Watts & Williamson, 2015; Perfecto et al., 2009).

The Food and Agriculture Organization (FAO) of the United Nations characterizes organic agriculture as an integrated system that promotes the health of all biological systems/ecosystems by eliminating synthetic chemicals and inputs (Food and Agriculture Organization [FAO], 2020). The U.S. Department of Agriculture (USDA) regulates organic agriculture in the U.S. where certified farmers are guided by mandated standards to avoid synthetic agrichemicals, sewage sludge, and radiated or bioengineered inputs (U.S. Department of Agriculture [USDA], 2021). Conservation agriculture promotes three core principles of farming: minimal mechanical disturbance of the soil; enhancing soil fertility and protecting the soil with cover crops and organic matter such as green manure; and planting a diverse array of crops that can increase nutrient cycling in soil and promulgate biodiversity (FAO, n.d.).

Dry Farming and Orphan Crops

Erratic weather patterns are now commonplace where many parts of the world are facing exceptional drought and heatwaves (United Nations Convention To Combat Desertification, 2022). Water is both a precious and undervalued resource that our livelihoods and survival depend upon (United Nations Convention To Combat Desertification; Hawkins, 2022). An agriculture practice that is largely overlooked in the U.S. but is used in other regions of the world is dry farming (Garrett 2019; Hawkins, 2022). Dry farming is also known as dryland farming or rainfed farming (Garrett 2019). Dry farming focuses on enhancing the water-building capacity of soil amidst a lack of irrigation where the soil retains moisture that was received from rainfall (Garrett 2019). Growing food during the dry season requires a deep understanding of the respective soil and soil-building techniques (using cover crops, green manure, and other organic mulches) (Garrett 2019).

Orphan crops can be described as underutilized local crops that aren't traded in the mainstream food system (Global Pulse Confederation, n.d.; Hawkins, 2022; Tadele, 2019). Orphan crops are culturally important, contribute to food security—and contribute to resilient ecosystems and agrobiodiversity (Global Pulse Confederation, n.d.; Borelli et al., 2020). As the industrialized food system has grossly limited the availability of crops and subsequently nutrients, orphan foods are particularly promising in vulnerable communities (Borelli et al., 2020). Many of these diverse plant foods are nitrogen-fixers; nutrient-dense; affordable—and can survive amidst climate stressors (Global Pulse Confederation, n.d.; Tadele, 2019). Orphan crops may include cereals, pseudo-cereals, legumes, pulses, vegetables, fruits, oilseeds, roots, tubers, and spices (Global Pulse Confederation, n.d.; Tadele, 2019; Guruani et al., 2021). The African Orphan Crop Consortium prioritizes over 100 Indigenous and traditional orphan crops that hold promise such as West African okra, vine spinach, papaya, horned melon, African mangosteen, jujube, sour plum, and macadamia nut (African Orphan Crops Consortium, n.d., Hawkins, 2022).

A study of the types and uses of wild plants by the Balete tribe of Ramotswa Village in Botswana found that over 50 wild plants are consumed in traditional dishes that includes fruits, seeds, nuts, root vegetables, tubers, and leafy vegetables (Bultosa et al. 2020; Hawkins, 2022). Varying parts of the plant such as the bark, rind, skins, gums, or roots may be used in cooking (Bultosa et al., 2020). Overall, orphan crops are considered a key solution in closing the gaps in food and nutrient security and can lend to positive planetary health outcomes if carried forth with integrity.

Spotlight: Agroecological Farming Practices and Achieving Nutrient Security

An inspiring example of a regenerative farming system that averts irrigation and maximizes planetary health comes from the central Himalayan region of India with a technique called barahnaja (Guruani et al., 2021; Hawkins, 2022). In this mixed-crop "companion-planting" system, plants nourish and support each other and are harvested at different times of the year, increasing food security (Guruani et al., 2021). The name barahnaja translates to the twelve crops that are grown together which includes pulses, spices, fibers, and vegetables (Guruani et al., 2021; Hawkins, 2022). The merits of this system include: the impressive nutrient density of the harvest; enhanced soil fertility; intimate knowledge of the terrain and living systems that dictates what is grown and where and when it is planted; crop rotation and letting the land rest; preventing erosion; organic farming practices; and recognizing the long-standing evolutionary relationships between plant roots and soil biodiversity (Guruani et al., 2021; Hawkins, 2022). Common crops used include finger millet, amaranth, buckwheat, sorghum, maize, kidney bean, blackgram (black lentils), horsegram (lentil), local soybeans, ricebean (legume), pigeon pea (legume), cow pea (legume), perilla (green plant), sesame, hemp, cucumber, and cleome (a leafy green) (Guruani et al., 2021).

The Indigenous Rukai peoples of southern Taiwan (Wutai Village) utilize traditional ecological knowledge to cultivate over 78 local varieties of 26 species of food crops amidst mountainous terrain (Ba et al. 2018; Hawkins, 2022). Local strains of basic food crops are cultivated: cereals and grains, legumes, oil crops, roots & tubers, fruits, vegetables, and sugar plants (Ba et al., 2018). Similar to barahanja farming techniques, the Rukai bolster planetary health in this unique bioregion by utilizing a mixed-crop rotation as well as limiting the use of arable land, enhancing soil fertility, saving seeds, maximizing nutrition security, and strategically utilizing plant and tree roots to prevent erosion (Ba et al., 2018). In fact, Wutai Village is one of the Indigenous cultures of Taiwan where traditional farming is the mainstay of the diet (Ba et al., 2018). Furthermore, the entire month of August celebrates the harvest which commences the tribal year and is followed by a ceremony called "capi" that predicts the conditions of the forthcoming harvest (Ba et al., 2018). The agricultural practices of both the Rukai peoples and barahnaja in the central Himalayan region of India demonstrate what the International Panel of Experts on Sustainable Food Systems (2016) recommends: the shift from uniformity in agriculture to highly diverse farming systems that support ecosystems, promulgates biodiversity—and offers long-term fertility that sustains livelihoods.

Anchored in Native Plant Ecosystems

Indigenous plants and the cuisines they create delineate important histories and stories as they evolved to local conditions dictated by the geology, soil, weather, and microclimate to maximize planetary health (Tallamy, 2019). As an example, the native pecan tree in Texas (U.S.) was traced to pre-historic times where Indigenous peoples consumed them in a variety of ways (Evans 1995; Travis, 2009; Hawkins, 2022). During the great economic depression in the U.S., between twelve and fifteen thousand people hand-shelled pecans for employment (Evans, 1995). The commercialization of pecans led to the mass production of cultivars, mechanization, and irrigation (Evans, 1995). Acorns, another native nut, have long played an important role for Indigenous peoples of the U.S. as a staple food (Olson, 2023).

The California Native Garden Foundation offers a model anchoring the food system in native ecosystems that robustly support planetary health (California Native Garden Foundation, n.d). As the California Native Garden Foundation is based in northern California, native grass and oak woodland plant communities comprise this ecosystem (California Native Garden Foundation, n.d). Native plants generally eliminate the resource utilization and inputs associated with mainstream/conventional gardening. Restoring the indigenous flora of local ecosystems also offers an array of nutritious foods and depending on the respective bioregion, may include foods such as acorns, amaranth, red huckleberry, evergreen huckleberry, blue elderberry, black elderberry, rooreh (Miner's lettuce), chia seeds, sages, salal berry, quail bush, mallow root, currants, thimbleberry, lemonade berry, tanoak mushrooms, manzanita berry, salmonberry, oso berries, currants, fiddlehead ferns, saskatoon berries, morel, spruce tips, strawberry, juniper berries, blackberry, pine nuts, pecans, hazelnuts, walnuts, yampah, flax, crab apple, and nettle (California Native Garden Foundation, n.d.; Olson, 2023). Using nettle as an example, it's an extremely versatile herbaceous plant used for soups, noodles, sauces, tea, dips, pesto, and baked goods (Kregiel et al., 2018; Hawkins, 2014). Nettle is nutrient-dense and naturally low in calories (Kregiel et al., 2018; Hawkins, 2014).

After reestablishing or bolstering native ecosystems, the Food Pyramid for the 21st Century recommends growing perennial foods as they generally require fewer inputs and grow over many seasons (California Native Garden Foundation, n.d). Thereafter, drought tolerant plants that are compatible with local soils are recommended, followed by nitrogen-fixing plants—and then foods with high nutrient-density. Lastly, "comfort foods" are grown as desired because they harbor a special sentiment (California Native Garden Foundation, n.d). Altogether, Indigenous flora are singular, beautiful, and keenly support biodiversity; many are medically important, too (Kregiel et al., 2018; Tallamy, 2019). Restoring Indigenous ecosystems is an essential solution to our planetary health crises and will be further discussed in Chapter 11.

Seed Sovereignty

Saving seeds and seed sovereignty are also key components of food sovereignty and regenerative agroecology (Navdayna, n.d.). Using the philosophy of "Vasudhaiv Kutumbakan" which translates to "The Earth as One Family", Navdanya is a collective in India of over 6.5 million farmers with 111 community seedbanks in 17 of India's 28 states (Navdanya, 2016). The word Navdanya translates to "nine seeds", symbolizing that Indigenous seeds are a life-giving entity; by saving and using Indigenous seeds, traditional knowledge is used to achieve food sovereignty, food security—and positive planetary health outcomes (Navdanya, 2016).

In the U.S., Bill McDorman, a longtime professional seed-saving expert, shares that seed saving helps teach us to take care of ourselves (McDorman & Thomas, 2019). Seeds are linked to the survival of our food system, and connect us to broader environmental cycles that we are generally disconnected from (McDorman & Thomas, 2019). McDorman believes in the transformative nature of saving seeds—especially when united with others where we become increasingly powerful in creating planetary resilience (McDorman & Thomas, 2019).

NOTE

1 Food and Agriculture Organization. (2024, March 11). *One Health day.* United Nations. www.fao.org/one-health/resources/events/events-detail/ohd…20slogan%20for%20One%20Health,protect%20the%20health%20of%20all.

REFERENCES

African Orphan Crops Consortium (n.d.). *Meet the crops.* https://africanorphancrops.org/meet-the-crops/

Altieri, M.A., & Nicholls, C.I. (2020). Agroecology: Challenges and opportunities for farming in the Anthropocene. *International Journal of Agriculture and Natural Resources*, *47*(3), 204–215. https://doi.org/10.7764/ijanr.v47i3.2281

American Public Health Association (APHA). (2017). *Advancing a 'One Health' approach to promote health at the human-animal-environment interface.* Policy Number: 201712. https://apha.org/policies-and-advocacy/public-health-policy-st…h%20is%20defined%20as,%5D%20While%20the%20human%2C%20animal%2C

American Public Health Association (APHA). (2019, November 5). *Precautionary moratorium on new and expanding concentrated animal feeding operations.* Policy 20194. https://apha.org/policies-and-advocacy/public-health-policy-statements…oratorium-on-new-and-expanding-concentrated-animal-feeding-operations

Ba, Q-X., Lu, D., Kuo, W.H., & Lai, P. (2018). Traditional farming and sustainable development of an Indigenous community in the mountain area—A case study of Wutai Village in Taiwan. *Sustainability*, *10*(3370), 1–16. https://doi.org/10.3390/su10103370

Badgley, G., Freeman, J., Hamman, J.J., Haya, B., Trugman, A.T., Anderegg, W.R., & Cullenward, D. (2022). Systematic over-crediting in California's forest carbon offsets program. *Global Change Biology*, *28*, 1433–1445. https://doi.org/10.1111/gcb.15943

Barbier, E.B., Georgiou, I.Y., Enchelmeyer, B., & Reed, D.J. (2013). The value of wetlands in protecting Southeast Louisiana from hurricane storm surges. *PLoS One*, 8(3), e58715, 1–6. https://doi.org/10.1371/journal.pone.0058715

Battocletti, V., Enriques, L., & Romano, A. (2023). *The voluntary carbon market: Market failures and policy implications.* European Corporate Governance Institute. https://papers.ssrn.com/sol3/papers.cfm?abstract_id=4380899

Booker, C. (2024, September 25). *Booker, Adams, McGovern introduce climate-smart farm conversion bill.* [Press release]. www.booker.senate.gov/news/press/booker-adams-mcgovern-introduce-climate-smart-farm-conversion-bill

Borelli, T., Hunter, D., Padulosi, S., Amaya, N., Meldrum, G., Beltrame, D.M.D, Samarasinghy, G., Wasike, V.W, Guner, B., Tan, A., Dembele, Y.K., Lochetti, G., Sidibe, A., & Tartanac, F. (2020). Local solutions for sustainable food systems: The contribution of orphan crops and wild edible species. *Agronomy*, *10*(231), 1–25. https://doi.org/10.3390/agronomy10020231

Bultosa, G., Molapisi, M., Tselaesele, N., Kobue-Lekalake, R., Haki, G.D., Makhabu, S., Sekwati-Monang, B., Seifu, E., & Nthoiwa, G.P. (2020). Plant-based traditional foods and beverages of Ramotswa Village, Botswana. *Journal of Ethnic Foods*, *7*(1), 1–15.

California Native Garden Foundation. (n.d.). *California Native Garden Foundation's food pyramid for the 21st century.* https://cngf.org/food-pyramid-for-the-21st-century/

Centers for Disease Control and Prevention (CDC). (2023, September 28). *One Health basics.* www.cdc.gov/onehealth/basics/index.html

Center for Progressive Reform. (2005, September). *An unnatural disaster: The aftermath of Hurricane Katrina.* Publication #512. Center for Progressive Reform. https://digitalcommons.law.ggu.edu/cgi/viewcontent.cgi?article=1044&context=pubs

Evans, A.S. (1995). *Pecan Industry: Handbook of Texas Online.* Texas State Historical Association. www.tshaonline.org/handbook/entries/pecan-industry.

Food and Agriculture Organization. (2020). *Introduction to organic agriculture.* https://teca.apps.fao.org/en/technologies/8359/

Food and Agriculture Organization. (n.d.). *Conservation agriculture.* https://openknowledge.fao.org/server/api/core/bitstreams/024e17be-9fad-4556-be94-a8e2f229023d/content

Garrett, A. (2019). *Intro to dry farming organic vegetables.* Oregon State University Extension Service. EM 9229. Oregon State University. https://extension.oregonstate.edu/sites/extd8/files/documents/em9229.pdf

Gibbons, L.V. (2020). Regenerative—The new sustainable? *Sustainability*, *12*(5483), 1–19. https://doi.org/10.3390/su12135483

Global Pulse Confederation. (n.d.). *Orphan crops: A solution to global food system risks and an investment opportunity for future agricultural research.* www.globalpulses.com/files/knowledge_center/20181108181907_factsheet-orphan_crops.pdf

Gururani, K., Salej, S., Kuman, A., Joshi, D.C., Pandey, D., & Sharma, A.R. (2021). Mainstreaming *Barahnaja* cultivation for food and nutritional security in the

Himalayan Region. *Biodiversity and Conservation*, *30*, 551–574. https://doi.org/10.1007/s10531-021-02123-9

Hawkins, I.W. (2014). Food science and the natural environment. In S. Edelstein (Ed.), *Food science, an ecological approach*. Jones and Bartlett.

Hawkins, I.W. (2022, October 22). Food professionals for planetary health [Conference paper]. *International Conference on Chinese Food Culture: Biological and Ecological Perspectives*. Chiang Ching-kuo Presidential Library, Taipei City, Taiwan.

Intergovernmental Science-Policy Platform on Biodiversity and Ecosystem Services. (n.d.). *Conventional agricultural*. www.ipbes.net/glossary-tag/conventional-agricultural

International Panel of Experts on Sustainable Food. (2016). *From uniformity to diversity: A paradigm shift from industrial agriculture to diversified agroecological systems*. International Panel of Experts on Sustainable Food systems. www.ipes-food.org/_img/upload/files/UniformityToDiversity_FULL.pdf

International Union for the Conservation of Nature. (2020). Global Standard for Nature-based Solutions. A user-friendly framework for the verification, design and scaling up of NbS. https://doi.org/10.2305/IUCN.CH.2020.08.en

Kregiel, D., Pawlikowska, E., & Antolak, H. (2018). *Urtica* spp.: Ordinary plants with extraordinary properties. *Molecules*, *23*(1664), 1–21.

McDorman, B., & Thomas., S. (2019). The importance of saving seeds. In Hawkins, I.W. (Ed.). *Promoting biodiversity in food systems* (pp. 189–202). CRC Press.

Melanidis, M.S., & Hagerman, S. (2022). Competing narratives of nature-based solutions: Leveraging the power of nature or dangerous distraction? *Environmental Science and Policy*, *132*, 273–281. https://doi.org/10.1016/j.envsci.2022.02.028

Navdanya. (n.d.). *Biodiversity, agroecology regenerative organic agriculture: Seed sovereignty & food sovereignty*. www.navdanya.org/plugins/content/pdf_embed/assets/viewer/pdfjs/web/viewer.html?file=%2Fimages%2Fenglish-flip-chart.pdf

Navdanya. (2016). *Navdanya: An overview*. www.navdanya.org/component/content/article?id=7

Mercado, G., Wild, T., Hernandez-Garcia, Baptista, M.D., van Lierop, M., Bina, O., Inch, A. Sang, A.O., Buijs, A., Dobbs, C., Vasquez, A., van der Jagt, A., Salbitano, F., Falanga, R., Amaya-Espinel, J.D., Pereira, M.D., & Randrup, T.B. (2024). Supporting nature-based solutions via nature-based thinking across European and Latin American cities. *Ambio*, *53*, 79–94. https://doi.org/10.1007/s13280-023-01920-6

National Centers for Environmental Information. (2024, April 8). *Costliest U.S. tropical cyclones*. www.ncei.noaa.gov/access/billions/dcmi.pdf

Nitzan, D., Andreuzza, B.N., & Chattopadhyay, D. (2023). The food systems, One Health, and resilience (FOR) approach—Led by the FOR-Runners. *Sustainability*, *15*, 1–13. https://doi.org/10.3390/su151813889

Olson, S.V. (2023). *Chimi Nu'am: Native California foodways*. Heyday.

One Health High-Level Expert Panel (OHHLEP), Adisasmito, W.B., Almuhairi, S., Behravesh, C.B., Bilivogui, P., Bukachi, S.A., Casas, N., ... & Zhou, L. (2022). One Health: A new definition for a sustainable and healthy future. *PLoS Pathoggens*, *18*(6), 1–4. https://doi.org/10.1371/journal.ppat.1010537

Perfecto, I., Vandermeer, J., & Wright, A. (2009). *Nature's matrix: Linking agriculture, conservation and food sovereignty*. Earthscan.

Redvers, N., Celidwen, Y., Schultz, Cl, Horn, O., Githaiga, C., Vera, M., Perdristat, M., Plume, L.M., Kobei, D., Kain, M.C., Poelina, A., Rojas, J.N., & Blondin, B. (2022). The determinants of planetary health: An Indigenous consensus perspective. *Lancet Planetary Health*, *6*, e156–e163.

Redvers, N., Poelina, A., Schultz, C., Kobei, D.M., Githaiga, C., Perdrisat, M., ... & Blondin, B.S. (2020). Indigenous natural and first law in planetary health. *Challenges*, *11*(29), 1–12. https://doi.org/10.3390/challe11020029

Reed, G., Poelina, A., Schultz, C., Kobei, D., Githaiga, C., Pedrisat, M., Prince, D., & Blondin, B. (2024). There is no word for 'nature' in our language: Rethinking nature-based solutions from the perspective of Indigenous Peoples located in Canada. *Climatic Change*, *177*(32), 1–24. https://doi.org/10.1007/s10584-024-03682-w

Smith, N. (2005, June 11). *There's no such thing as a natural disaster*. Social Science Research Council. https://items.ssrc.org/understanding-katrina/theres-no-such-thing-as-a-natural-disaster/

Tadele, Z. (2019). Orphan crops: Their importance and the urgency of improvement. *Planta*, *250*, 677–694. https://doi.org/10.1007/s00425-019-03210-6

Tallamy, D.W. (2019). *Nature's best hope: A new approach to conservation that starts in your yard*. Timber Press.

Travis, D. (2009, May 19). *Native pecan tree*. El Camino Real Chapter of the Texas Master Naturalist Program Milam County. https://txmn.org/elcamino/files/2010/03/Native-Pecan-Tree.pdf

U.S. Department of Agriculture (USDA). (2021). *What's behind the organic seal?* www.ams.usda.gov/sites/default/files/media/NOP_BehindTheSeal.pdf

U.S. Environmental Protection Agency (EPA). (n.d.). *Regulatory definition of large CAFOs, medium CAFO, and small CAFOs*. www3.epa.gov/npdes/pubs/sector_table.pdf

United Nations Convention to Combat Desertification. (2022). *Drought in numbers 2022: Restoration for readiness and resilience*. www.unccd.int/resources/publications/drought-numbers

Watts, M., & Williamson, S. (2015). *Replacing chemicals with biology: Phasing out highly hazardous pesticides with agroecology*. Pesticide Action Network Asia and the Pacific. https://saicmknowledge.org/sites/default/files/resources/Replacing%20Chemicals%20with%20Biology.pdf

West, T.A.P., Wunder, S., Sills, E.O., Borner, J., Rifai, S.W., Neidermeier, A.N., Frey, G.P., & Kontoleon, A. (2023). Action needed to make carbon offsets from forest conservation work for climate change mitigation. *Science*, *381*(6660), 873–877. https://doi.org/10.1126/science.ade3535

World Health Organization, Food and Agriculture Organization of the United Nations, United Nations Environment Programme and World Organisation for Animal Health. (2023). *A guide to implementing the One Health Joint Plan of Action at national level*. https://iris.who.int/bitstream/handle/10665/374825/9789240082069-eng.pdf?sequence=1

Whole Plant Foods Nutrition for Personal and Planetary Resilience

9

"Plants are the foundation of all life on the planet. They are also the food of all life".

Ron Vanderhoff[1], Botanist, Horticulturalist, Photographer

OVERVIEW

The Global Network for the Right to Food and Nutrition (2020) writes, "Nowhere else is the interconnectedness between humans and nature more evident than in food. And yet we have been swept up in a delusion of separation, which today is at the root of a deep ecological crisis". This statement aptly describes the state of the mainstream food system and planetary health. At the individual level, there are a myriad of ways to improve health outcomes with diet and lifestyle behaviors that simultaneously offer a multitude of co-benefits. As such, this chapter reinforces the merits of whole plant foods nutrition and the associated co-benefits of nutrient security, reversing and preventing chronic disease—and stresses the importance of preventing food

DOI: 10.1201/9781003153528-11

waste and composting. Ultimately, whole plant foods nutrition is an invaluable tool in advancing planetary health.

WHOLE PLANT FOODS NUTRITION FOR PLANETARY HEALTH

Scientists across an array of disciplines have demonstrated that we will not achieve climate mitigation goals without shifting broadscale dietary patterns away from animal products (Intergovernmental Panel on Climate Change [IPCC], 2019). During the first press conference of the Intergovernmental Science-Policy Platform on Biodiversity and Ecosystem Services (IPBES), leaders declared the need for transformative change across the globe that included "changing the way we eat" to minimize the risk to a million species in peril (Intergovernmental Science-Policy Platform on Biodiversity and Ecosystem Services [IPBES], 2019a). The United Nations Convention To Combat Desertification's (UNCCD) Executive Secretary Ibrahim Thiaw (United Nations Convention To Combat Desertification [UNCCD], 2022) declared (Hawkins, 2022):

> We must build and rebuild our landscapes better, mimicking nature wherever possible and creating functional ecological systems. Restoration is not enough, however. We need to protect and manage lands through improved consumption and production practices. On the agriculture side, this means sustainable and efficient management techniques that grow more food on less land and with less water. On the consumption side, this means changing our relationships with food, fodder and fiber, moving toward plant-based diets, reducing or stopping the consumption of animals.

The Intergovernmental Panel on Climate Change (IPCC) recommends the shift to plant-based diets for the greatest reduction in greenhouse gases emissions as noted in Figure 9.1 where the vegan dietary pattern (containing no animal foods) offers the best outcomes compared to other dietary patterns such as vegetarian, flexitarian, pescetarian, and Mediterranean (IPCC, 2019).

The EAT-Lancet Commission (EAT) assessment of food and planetary health called for a "Great Food Transformation" where unprecedented changes in food production are recommended to improve food security, human health—and to increase the likelihood of maintaining a safe operating space within the Earth System (Willett et al., 2019). To achieve this, their "Planetary Health

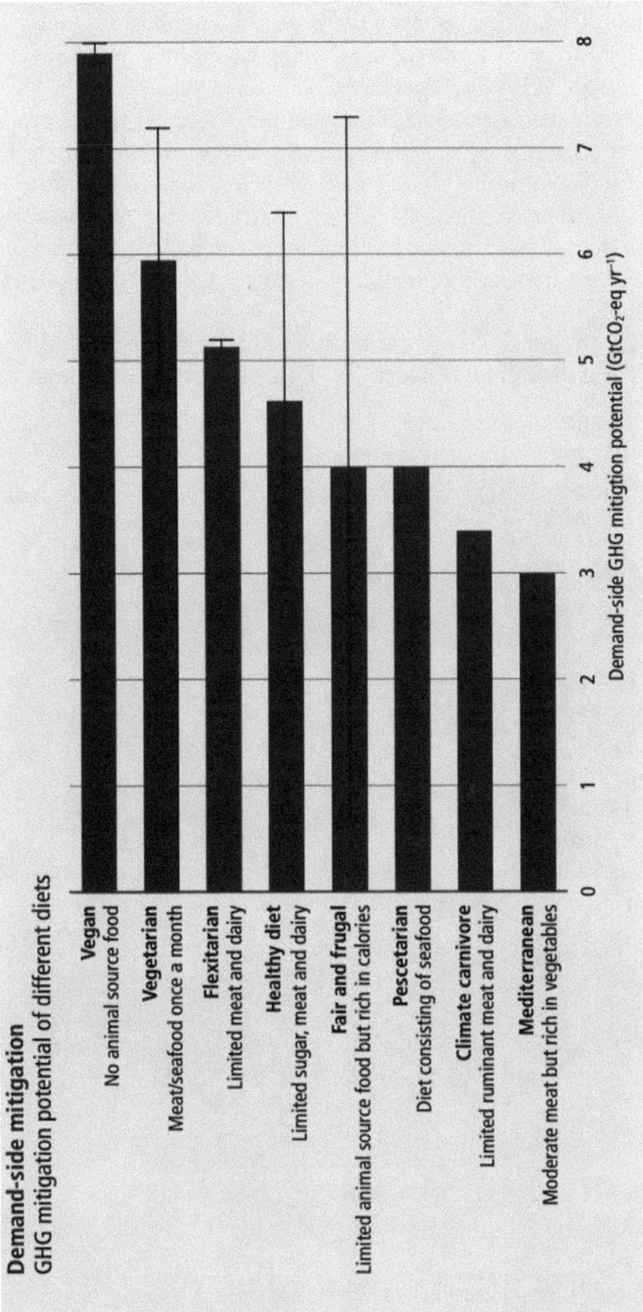

FIGURE 9.1 IPCC 2019 Special Report, Technical Mitigation Potential of Changing Diets by 2050 According to a Range of Scenarios Examined in the Literature.

Diet" calls for the greatest percentage of the daily dietary pattern to come from fruits, vegetables, whole grains, plant proteins, and unsaturated plant oils with limited amounts of animal proteins, dairy, added sugars, and starchy vegetables as seen in Figure 9.2 (EAT-Lancet Commission [EAT], n.d.).

A cross-sectional assessment of the land use, water use, and greenhouse gas emissions (GHGE) associated with U.S. dietary patterns (the standard diet, a healthy diet, a healthy Mediterranean diet, a healthy vegetarian, or a healthy vegan diet) revealed that the healthy vegan diet had the lowest impact where omnivorous diets containing red meat had the highest (Jennings et al., 2023). The Mediterranean diet contributed to the highest GHGE (Jennings et al., 2023).

A prospective analysis of the health and environmental impacts of a large U.S.-based cohort between 1991–2017 found that animal-based

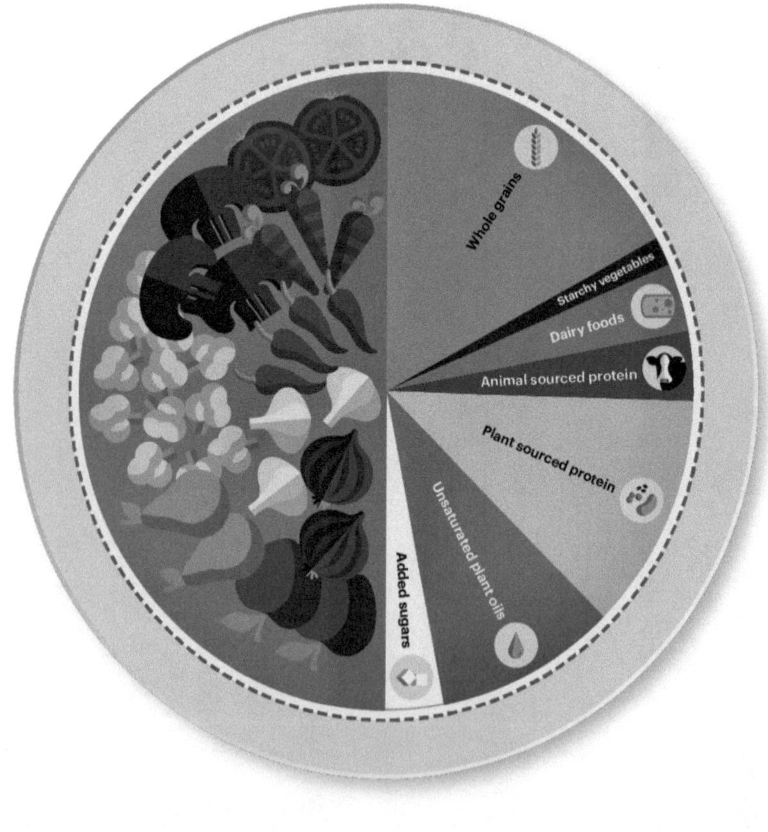

FIGURE 9.2 EAT-Lancet Commission "Planetary Health Plate" (eatforum.org/eat-lancet-commission).

dietary patterns contributed the most to greenhouse gas emissions, cropland use, and irrigation water—particularly diets containing red and processed meats, dairy, poultry, and fish, respectively (Musicus et al., 2022). Higher Plant-based Diet Index (PDI) scores (a dietary indices associated with the consumption of plant-based foods) was associated with a decreased risk of cardiovascular disease along with decreased greenhouse gas emissions, nitrogen, cropland use, and irrigation however, an unhealthy PDI score (a dietary indices associated with plant-based foods that are higher in refined products and added sugars) was associated with the increased risk of cardiovascular disease as well as cropland and fertilizer use, respectively (Musicus et al., 2022). Both higher Alternative Healthy Eating Index (AHEI) scores (which includes more fruits and vegetables, long chain fatty acids and polyunsaturated fats) and higher healthy PDI scores experienced a lower risk of chronic diseases, lower greenhouse gas emissions—and decreased cropland, irrigation, and fertilizer use, respectively (Musicus et al., 2022).

WHOLE PLANT FOODS NUTRITION, FOOD WASTE, AND FOOD SECURITY

Globally, Americans waste the most food (Recycle Track Systems, n.d.). In developing countries, food is lost due to a lack of proper infrastructure and storage capabilities (Guo et al., 2023). When food is wasted, not only are the nutrients wasted—but all inputs along the way are wasted such as water, labor, energy, etc. (Conrad et al., 2018). Food loss and food waste contribute to air pollution and GHGE, too (Guo et al., 2023). Conrad et al. (2018) calculated that Americans waste nearly one pound of food daily, with produce accounting for the largest percentage of waste. Higher quality diets contained more produce and thus accounted for more food waste than lower quality diets, respectively. These authors concluded, "...increasing consumers' knowledge about how to prepare and store fruits and vegetables will be an essential component to reducing food waste" (Conrad et al., 2018).

Food is also left in the fields on the farm (Baker et al., 2019). An assessment of food waste at midsize and large conventional farms in central and northern California found that 31% of the edible harvest was left in the field; another 2.4% of crops are not harvested whatsoever (Baker et al., 2019). Globally, over 15% of food produced is wasted on farms where the goal moving forward is to cut food waste in half by 2030 (World Wildlife Federation United Kingdom, 2021). In the U.S., the greatest percentage of waste in landfills comes from food, where it produces methane, a powerful greenhouse gas that was discussed in Chapter Four (U.S. Environmental Protection Agency, 2020; Nesser et al., 2019).

Animal-based foods are an inefficient way to achieve nutrient intake (Shepon et al., 2018; Siddhartha, 2018). We could feed nearly all Americans if animal-based agriculture shifted to plant-based agriculture and plant-based dietary patterns (Shepon et al., 2018). A study of 150 countries found that shifting to a predominantly plant-based dietary pattern would increase nutrient intake, decrease impact to the natural environment, and decrease food-related morbidity and mortality (Springmann et al., 2018). Additionally, if crops grown for animals and biofuels were instead grown for human consumption, an additional 4 billion people in the world could be fed (Cassidy et al., 2013).

WHOLE PLANT FOODS NUTRITION FOR PERSONAL HEALTH AND LIFESTYLE MEDICINE

The Academy of Nutrition and Dietetics (U.S.) position on vegetarian diets has long stated that well-planned diets that contain no animal products are not only safe at every stage of the lifecycle—but offer a multitude of health benefits in disease prevention and amelioration along with environmental care (Melina et al., 2016; Hawkins, 2020). Whole plant foods nutrition is associated with increased longevity and improved nutrient intake, too (Farmer et al., 2011 Fraser & Shavlik, 2001). Additionally, some researchers have dispelled the idea that the evolutionary development of the human brain was due to carnivory (Barr et al., 2022).

Lifestyle medicine focuses on the prevention of chronic disease where the majority of nutrition-related chronic diseases can be prevented (Bodai et al., 2018). The American College of Lifestyle Medicine (ACLM) delineates six foundational constructs to prevent, treat, and reverse nutrition-related chronic diseases: 1) whole plant-foods nutrition 2) physical activity 3) stress management 4) avoidance of risky substances 5) restorative sleep and 6) positive social connections (American College of Lifestyle Medicine [ACLM], 2023). The ACLM's "Whole Food, Plant-Based Plate" for adults is depicted in Figure 9.3 which is considered a "nutrition prescription" for treating and reversing chronic disease.

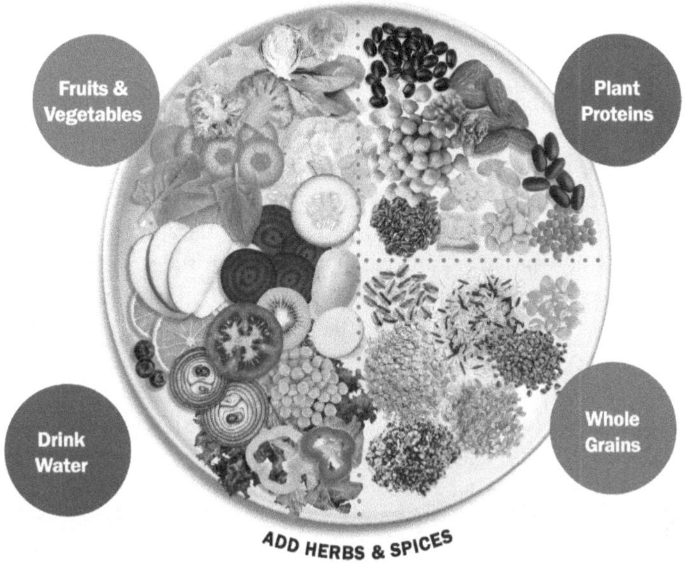

FIGURE 9.3 Whole Food, Plant-Based Plate, American College of Lifestyle Medicine.

FORGING NEW PATHS WITH WHOLE PLANT FOODS COOKERY

It's been said, "The road to health goes through the kitchen" (Palmer, 2019). The professions of dietetics and home economics have long asserted that cooking from scratch alongside prioritizing nutrition and family resource management is fundamental to creating healthy families, achieving an optimal quality of life, and protecting public health (American Association of Family & Consumer Sciences, 2024; Academy of Nutrition and Dietetics, 2024). As more food is eaten outside the home, cooking is now an underutilized skill although it's an essential strategy in safeguarding health and mitigating impact to the natural environment (Palmer, 2019). Cooking from scratch allows an individual to improve dietary quality by controlling the ingredients procured along with cooking methods, portion sizes—and avoiding packaging (Palmer, 2019; Wolfson & Bleich, 2014). Although ready-to-make home meal kits have become popular in recent years, the negative environmental impact of plastic food containers is staggering. With ultra-processed convenience foods now commonplace, cooking fresh whole plant foods is a powerful, health-promoting tool.

Registered Dietitian Nutritionists (RDN's) are the board-certified healthcare provider solely trained in the application of food, nutrition, and dietetics across the lifecycle (Hawkins et al., 2019). Overwhelmingly, RDN's in the U.S. are concerned with climate change and over 90% of accredited dietetics program directors surveyed agreed that vegetarian and vegan nutrition should be taught—and also perceive that student attitudes towards vegetarian and vegan nutrition are favorable (Hawkins et al., 2015; Hawkins et al., 2019). Furthermore, vegetarian and vegan RDN's in the U.S. were significantly more likely than non-vegetarian and non-vegan dietitians to engage in actions that used diet as a climate mitigation strategy (Hawkins et al., 2015).

Those accredited dietetics programs in the U.S. that teach hands-on vegan and vegetarian cookery skills describe experiential learning that included but was not limited to assigning students to consume a vegetarian or a vegan diet for five days and reflect upon the experience; recipe development; sensory evaluations; hosting tasting events; food demonstrations; patient/client counseling simulation; field trips; and emphasizing multicultural cuisine and ethnic cooking traditions (Hawkins & Mangels, 2021).

As physicians are gatekeepers on the healthcare team, those that promote lifestyle medicine must demonstrate effective leadership skills to create new pathways amidst healthcare systems that offer costly and invasive medical interventions to treat nutrition-related chronic diseases (Stancic, 2019). In practice, the lifestyle medicine team includes but is not limited to physicians, dietitians, behavioral therapists, and exercise coaches (Bodai et al., 2018). In recent years, milestones have been achieved with integrating whole plant foods nutrition into patient care. The Montefiore Cardiac Wellness Program in the Bronx (New York) offers a physician and dietitian-led wellness program demonstrating how a 100% plant-based diet and lifestyle medicine maximizes health outcomes (Montefiore Medical Center, 2024). The University of Florida Shands Hospital offers a plant-based menu for inpatients, outpatients, and employees (Saldivar et al., 2022). Other notable hospital programs are found in Tampa and Denver (Aggarwal et al., 2019). New York City has the largest municipal hospital system in the U.S. where all eleven public hospitals offer culturally-diverse plant-based meals (The City of New York, 2024). In the outpatient setting, the Barnard Medical Center in Washington D.C. utilizes whole plant foods nutrition and lifestyle behaviors to reverse nutrition-related chronic disease, often resulting in the elimination of prescription medicines (Physicians Committee for Responsible Medicine, 2024).

California's Senate Bill 1138 requires healthcare facilities and prisons to make wholesome, plant-based meals available to patients and inmates (State of California, 2018). The American Medical Association's 2021 resolution

H-150.949 "Healthful Food Options in Healthcare Facilities" calls on all healthcare facilities to provide accessible and healthy food options including plant-based meals (American Medical Association, 2021).

UNIVERSITY TEACHING KITCHENS AND CULINARY MEDICINE PROGRAMS

Decreased cooking amongst the general public as well as the rampant increase in nutrition-related chronic disease has informed the advent of culinary medicine programs and university teaching kitchens. University teaching kitchens may offer cooking and nutrition education programs to a wide range of populations including patients, students, and staff—as well as summer camps, too (Regents of the University of California, 2024; Oregon Health & Science University, 2025). Figure 9.4 represents a sign promoting root vegetable cookery offered by RDN's at Oregon Health & Science University (Portland, Oregon).

Additionally, the Culinary Institute of America and the Harvard School of Public Health teamed up to create the "Menus of Change" program where "plant-forward" and "veg-centric" menus embrace nutrient-dense fruits, vegetables, whole grains, plant proteins, nuts, and legumes as the "centerpiece" of the meal; when meat is utilized, it should be considered a "condiment" (The Culinary Institute of America, 2024).

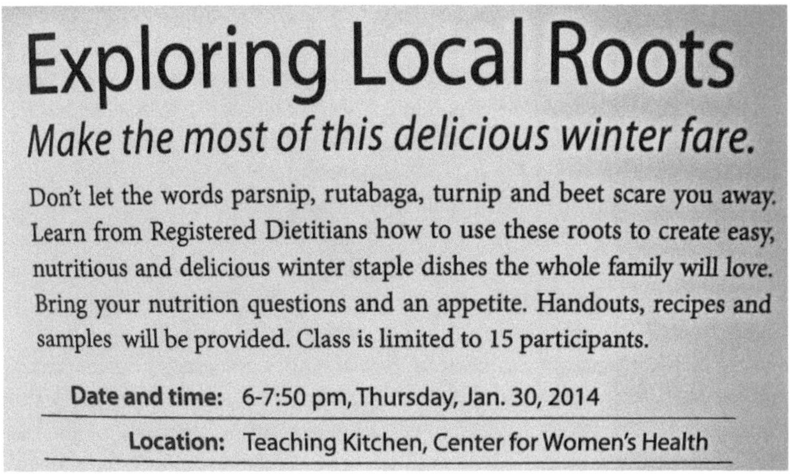

FIGURE 9.4 Whole Plant Foods Nutrition Cooking Class Sign at Oregon Health & Sciences University Women's Clinic in Portland, Oregon. Photo © Irana Hawkins.

Because of the escalating burden of chronic disease and the limited amount of nutrition knowledge practicing physicians have, medical schools are increasingly recognizing the importance of nutrition and lifestyle behaviors by creating "culinary medicine" curriculums (Tan et al., 2022). Culinary medicine programs have been described as health practitioner-led teaching interventions focused on combining culinary skills along with the health aspects of nutrition (Croxford et al., 2024). These experiential programs offer a hands-on approach to increase students' skills and knowledge and ultimately, the positive health outcomes of their future patients (Tan et al., 2022).

In addition to teaching cooking skills, some programs focus on closing the chasm of injustices and food insecurity. These "Food As Medicine" programs may be anchored in public health departments, universities, or teaching hospitals. These programs will be profiled in forthcoming chapters.

COMPOSTING WHOLE PLANT FOOD SCRAPS

Cooking whole plant foods from scratch results in inedible food scraps such as cores, peels, husks, and rinds. Although some municipalities in the U.S. accept food scraps in curbside green waste bins, these items can be composted at home where such nutrients from the Earth can be composted into a soil-like medium and delivered right back to the Earth. Home composting decreases greenhouse gas emissions, increases soil fertility, and promulgates biodiversity (Food and Agriculture Organization, 2015). Composting can be as easy as burying the compostable items into the soil; creating a compost trench or a compost pile; sheet mulching; or utilizing a compost bin or a worm bin (vermicomposting) (Oregon State University Extension, 2017).

Many compost bins readily available for sale are made of plastic and may be expensive. However, there are numerous alternatives to plastic compost bins that are easy to construct using wire, brick, wood, or cinder block (South Carolina Department of Public Health, 2018). Another alternative to a plastic bin is a galvanized steel bin and lid that resembles a trash can with holes in it. Simply placing compostable food items in this container and its accompanying tight-fitting lid outdoors (preferably on soil versus concrete so the drippings can nourish the Earth below) allows nature to take care of the rest. This compost bin is durable, space-efficient—and can eventually be recycled if necessary. Although some may choose to add red wiggler worms to facilitate the process, black soldier flies may appear independently and are excellent decomposers. Additionally, there are small compost bins designed for interior use that range from under-the-sink models to electric machines.

In closing, the scientific evidence demonstrates that whole plant foods nutrition is a necessary solution for improving food security, personal, and planetary health. This concept is increasingly embraced by healthcare professionals where innovative programs and pathways are being created. An additional co-benefit of whole plant foods nutrition includes returning inedible food scraps to the Earth as compost, another form of nutrition.

NOTE

1 Vanderhoof, R. (n.d.) Personal communication, February, 2024.

REFERENCES

Academy of Nutrition and Dietetics. (2024). *About the academy.* www.eatright.org/about-the-academy

Aggarwal, A., Grady, A., Desai, D., Hartog, K., Correa, L., Ostfeld, R.J., Freeman, A.M., McMacken, M., Gianos, E., Reddy, K., Batiste, C., Wenger, C., Blankstein, R., Williams, K., Allen, K., Seifried, R.M., Aspry, K., & Barnard, N.D. (2019). Successful implementation of healthful nutrition initiatives into hospitals. *American Journal of Medicine*, *133*, 19–25.

American Association of Family and Consumer Sciences. (2024). *About AAFCS.* www.aafcs.org/partnership/about-aafcs

American College of Lifestyle Medicine. (2023). *Six ways to take control of your health.* https://lifestylemedicine.org/wp-content/uploads/2023/06/Pillar-Booklet.pdf

American Medical Association. (2021). *Healthful food options in health care facilities H-150.949.* https://policysearch.ama-assn.org/policyfinder/detail/Healthy%20Food%20Options%20in%20Hospitals%20H-150.949?uri=%2FAMADoc%2FHOD.xml-0-627.xml

Baker, G.A., Gray, L.D., Harwood, M.J., Osland, T.J., & Tooley, J.B.C. (2019). On-farm food loss in northern and central California: Results of field survey measurements. *Resources, Conservation & Recycling*, *149*, 541–549. https://doi.org/10.1016/j.resconrec.2019.03.022

Barr, A.W., Pobiner, B., Rowan, J., Du, A., & Faith, J.T. (2022). No sustained increase in zooarchaeological evidence for carnivory after the appearance of *Homo erectus*. *Proceedings of the National Academy of Sciences*, *119*(5), e2115540119. https://doi.org/10.1073/pnas.2115540119

Bodai, B.I., Nakata, T.E., STAR Provider, CWFPBN, Wong, W.T., Clark, D.R., Lawenda, S., Tsou, C.; Liu, R., Shiue, L., Cooper, N., Rehbein, M., Ha, B.P., McKeirnan, A., Misquitta, R., Vij, P., Klonecke, A., Mejia, C.S., Dionysian, D., Hashmi, S., Greger, M., Stoll, S., & Campbell, T.M. (2018). Lifestyle medicine: A brief review of its dramatic impact on health and survival. *Permanente Journal*, *22*, 17–25. https://doi.org/10.7812/TPP/17-025

Cassidy, E.S., West, P., Gerber, J.S., & Foley, J.A. (2013). Redefining agricultural yields: from tonnes to people nourished per hectare. *Environmental Research Letters*, *8*, 1–8. https://doi.org/10.1088/1748-9326/8/3/034015

Conrad, Z., Niles, M.T., Neher, D.A., Roy, E.D., Tichenor, N.E., & Jahns, L. (2018). Relationship between food waste, diet quality, and environmental sustainability. *PLoS ONE*, *13*(4), e0195405. https://doi.org/10.1371/journal.pone.0195405

Croxford, S., Stirling, E., MacLaren, J., McWhorter, J.W., Frederick, L., & Thomas, O.W. (2024). Culinary medicine or culinary nutrition? Defining terms for use in education and practice. *Nutrients*, *16*(603), 1–12. https://doi.org/10.3390/nu16050603

EAT-Lancet Commission. (n.d.). *Adapted summary. Food in the Anthropocene: The EAT-Lancet commission on healthy diets from sustainable food systems.* https://eatforum.org/content/uploads/2019/07/EAT-Lancet_Commission_Summary_Report.pdf

Farmer, B., Larson, B.T., Fulgoni, V.L., Rainville, A.J., & Liepa, G.U. (2011). A vegetarian dietary pattern as a nutrient-dense approach to weight management: An analysis of the National Health and Nutrition Examination Survey 1999–2004. *Journal of the American Dietetic Association*, *111*(6), 819–827. https://doi.org/10.1016/j.jada.2011.03.012

Food and Agriculture Organization. (2015). *Composting: let's give the soil something back.* United Nations. www.fao.org/soils-2015/news/news-detail/en/c/280674/

Fraser, G.E., & Shavlik, D.J. (2001). Ten years of life: Is it a matter of choice? *Archives* of *Internal Medicine*, *161*(13), 1645–52. https://doi.org/10.1001/archinte.161.13.1645

Global Network for the Right to Food and Nutrition. (2020, October). *Overcoming ecological crises: Reconnecting food, nature and human rights.* www.fian.org/en/publication/article/right-to-food-and-nutrition-watch-2020-2630

Guo, Y., Tan., H., Zhang, L., Liu, G., Zhou, M., Vira, J. Hess, P.G., Liu, X., Paulot, F., & Liu, X. (2023). Global food loss and waste embodies unrecognized harms to air quality and biodiversity hotspots. *Nature Food*, *4*, 686–698.

Hawkins, I.W., & Mangels, A.R. (2021). Resources used and innovations in teaching vegetarian and vegan nutrition in accredited dietetics programs in the United States. *International Journal of Disease Reversal and Prevention*, *3*(2), 14–26. https://doi.org/10.22230/ijdrp.2021v3n2a231

Hawkins, I.W., Mangels, A.R., Goldman, R., & Wood, R.J. (2019). Dietetics program directors in the United States support teaching vegetarian and vegan nutrition and half connect vegetarian and vegan diets to environmental impact. *Frontiers in Nutrition*, *6*(123), 1–10. https://doi.org/10.3389/fnut.2019.00123

Hawkins, I.W., Balsam, A.L., & Goldman, R. (2015). A survey of Registered Dietitians' concern and actions regarding climate change in the United States. *Frontiers in Nutrition*, *2*(21), 1–8. https://doi.org/10.3389/fnut.2015.00021

Hawkins, I.W. (2022, October 22). Food professionals for planetary health [Conference paper]. *International Conference on Chinese Food Culture: Biological and Ecological Perspectives*. Chiang Ching-kuo Presidential Library, Taipei City, Taiwan.

Hawkins, I.W. (2020). *RDN resources for consumers: Plant-based diets to combat climate change and protect planetary health*. Vegetarian Nutrition Dietetics Practice Group, Academy of Nutrition and Dietetics (USA). https://higherlogicdownload.s3.amazonaws.com/THEACADEMY/859dd171-3982-43db8535-56c4fdc42b51/UploadedImages/VN/Documents/Resources/Climate_Change_Consumer_RDN_Consumer_V5.pdf

Intergovernmental Panel on Climate Change [IPCC]. (2019). *Summary for policymakers. In: Climate Change and land: An IPCC special report on climate change, desertification, land degradation, sustainable land management, food security, and greenhouse gas fluxes in terrestrial ecosystems*. www.ipcc.ch/site/assets/uploads/sites/4/2022/11/SRCCL_SPM.pdf

Intergovernmental Science-Policy Platform on Biodiversity and Ecosystem Services (IPBES). (2019a, May 6). Personal Communication.

Intergovernmental Science-Policy Platform on Biodiversity and Ecosystem Services (IPBES). (2019b). *Summary for policymakers of the global assessment report on biodiversity and ecosystem services of the Intergovernmental Science-Policy Platform on Biodiversity and Ecosystem Services*. https://doi.org/10.5281/zenodo.3553579

Jennings, R., Henderson, A.D., Phelps, A., Janda, K.M., & van den Berg, A.E. (2023). Five U.S. dietary patterns and their relationship to land use, water use, and greenhouse gas emissions: Implications for future food security. *Nutrients*, *15*(215), 1–15. https://doi.org/10.3390/nu15010215

Melina, V., Craig, W.R., & Levin, S. (2016). Position of the academy of nutrition and dietetics: Vegetarian diets. *Journal of the Academy of Nutrition and Dietetics*, *116*(12), 1970–1980.

Montefiore Medical Center. (2024). *Cardiac wellness program: Reversing heart disease with a whole food/plant-based diet*. www.montefiore.org/cardiacwellnessprogram

Musicus, A.A., Wang, D.D., Janiszewski, M., Eshel, G., Blondin, S.A., Willett, W., & Stampfer, M.J. (2022). Health and environmental impacts of plant-rich dietary patterns: A US prospective cohort study. *Lancet Planetary Health*, *6*, e892–e900.

Nesser, H., Jacob, D.J., Maasakkers, J.D., Lorente, A., Chen, Z., Lu, X., Shen, L. Qu, Z., Sulpriziol, M.P., Winter, M., Ma, S., Bloom, A.A., Worden, J.R., Stavins, R.N., & Randles, C.A. (2024). High-resolution US methane emissions inferred from an inversion of 2019 TROPOMI satellite data: contributions from individual states, urban areas, and landfills. *Atmospheric Chemistry and Physics*, *24*, 5069–5091. https://doi.org/10.5194/acp-24-5069-2024

Oregon Health & Science University. (2025). *Teaching kitchen*. www.ohsu.edu/march-wellness/teaching-kitchen

Oregon State University Extension. (2017). *Three methods for no-turn cold composting*. https://extension.oregonstate.edu/node/101001/printable/print

Palmer, S. (2019). The road to health goes through the kitchen. In Hawkins, I.W. (Ed.). *Promoting biodiversity in food systems* (pp. 247–261). CRC Press.

Physicians Committee for Responsible Medicine. (2024). *About the Barnard Medical Center*. www.pcrm.org/barnard-medical-center/about

Recycle Track Systems. (n.d.). *Food waste in America in 2024: Statistics + facts*. www.rts.com/resources/guides/food-waste-america/#:~:text…ood"%20for%20thought,120%20billion%20pounds%20—%20every%20year.

Regents of the University of California. (2024). *UCLA teaching kitchen*. https://fitwell.recreation.ucla.edu/facilities/teachingkitchen

Saldivar, B., Al-Turk, B., Brown, M., & Aggarwal, M. (2021). Successful incorporation of a plant-based menu into a large academic hospital. *American Journal of Lifestyle Medicine*, *16*(3), 311–317.

Shepon, A., Eshel, G., Noor, E., & Milo, R. (2018). The opportunity cost of animal based diets exceeds all food losses. *Proceedings of the National Academy of Sciences*, *115*(15), 3804–3809.

Siddhartha, K. (2019). Industrial agriculture, biodiversity, and planetary boundaries. In Hawkins, I.W. (Ed.)., *Promoting biodiversity in food systems* (pp. 29–54). CRC Press.

South Carolina Department of Health and Environmental Control. (2018). *Composting: recycling naturally – simple steps for starting at home*. https://des.sc.gov/sites/des/files/Library/OR-1705.pdf

Springmann, M., Wiebe, K., Mason-D'Croz, Sulser, T.B., Rayner, M., & Scarborough, P. (2018). Health and nutritional aspects of sustainable diet strategies and their association with environmental impacts: A global modelling analysis with country-level detail. *Lancet Planetary Health*, *2*(10), e451–e461. https://doi.org/10.1016/S2542-5196(18)30206-7

Stancic, S. (2019). Redefining medical practice with lifestyle medicine and environmental care. In Hawkins, I.W. (Ed.). *Promoting biodiversity in food systems* (pp. 283–291). CRC Press.

State of California. (2018). *Senate Bill 1138, Chapter 512*. https://leginfo.legislature.ca.gov/faces/billPdf.xhtml?bill_id=201720180SB1138&version=20170SB113894CHP

Tan, J., Atamanchuk, L., Rao, T., Sato, K., Crowley, J., & Ball, L. (2022). Exploring culinary medicine as a promising method of nutritional education in medical school: A scoping review. *BMC Medical Education*, 22(441), 1–24. https://doi.org/10.1186/s12909-022-03449-w

The Culinary Institute of America. (2024). *Plant-forward kitchen*. www.plantforwardkitchen.org/overview

The City of New York. (2024). *NYC Health + Hospitals now serving culturally-diverse plant-based meals as primary dinner option for inpatients at all of its 11 public hospitals*. [Press release]. www.nychealthandhospitals.org/pressrelease/nyc-health-hospitals…primary-dinner-option-for-inpatients-at-all-of-its-11-public-hospitals/

United Nations Convention to Combat Desertification. (2022). *Drought in numbers 2022: Restoration for readiness and resilience*. www.unccd.int/resources/publications/drought-numbers

U.S. Environmental Protection Agency. (2020). *Advancing sustainable materials management: 2018 Fact sheet*. www.epa.gov/sites/default/files/2020-11/documents/2018_ff_fact_sheet.pdf

Willett, W., Rockström, J., Loken, B., Springmann, M., Lang, T., Vermeulen, S., Garnett, T., Tilman, D., DeClerke, F., Wood, A., Jonell, M., Clark, M., Gordon, L.J., Fanzo, J., Hawkes, C., Zurayk, R., Rivera, J.A., De Vries, W., Sibanda, L.M.,… & Murray, C.J.L. (2019). Food in the Anthropocene: The EAT–*Lancet* commission on healthy diets from sustainable food systems. *Lancet*, *393*, 447–492.

Wolfson, J.A., & Bleich, S.N. (2014). Is cooking at home associated with better diet quality or weight-loss intention? *Public Health Nutrition*, *18*(8), 1397–1406.

World Wildlife Federation United Kingdom. (2021). *Driven to waste: The global impact of food loss and waste on farms.* https://files.worldwildlife.org/wwfcmsprod/files/Publication/file/5p58sxloyr_technical_report_wwf_farm_stage_food_loss_and_waste.pdf

Community Approaches to Nutrition Security and Equity

10

"Our program focuses on the health benefits of vegetables, fruits, beans, whole grains and other plant-based whole foods as the foundation for a healthy diet. This type of eating pattern is consistently associated with overall longevity and recommended for the prevention and management of many chronic diseases, including cancer, diabetes and cardiovascular disease."

Marianna Wetherhill[1], PhD, MPH, RDN, University of Oklahoma Culinary Medicine Program

OVERVIEW

When touring the Boston Medical Center rooftop garden in 2019 with longtime employee Latchman Hiralal, DTR, co-founder of the Preventive Food Pantry program and rooftop garden, he explained these entities were a part of creating, "the healthiest hospital community in the world". I replied, "Did you say the healthiest hospital community *in the world*?" to which he offered a resounding, "Yes!" It may appear that growing fresh produce on the

DOI: 10.1201/9781003153528-12

roof of a hospital and distributing it to a patient by way of a "prescription" only benefits the recipient. But when an institution strives to care for a community that has been underserved and overlooked by using nature in the form of fresh whole foods as a vehicle to nourish individuals, promote equity, and reverse chronic disease—there is now an opportunity to transform the community altogether. This chapter focuses on efforts that are moving communities towards equity and nutrition security with novel approaches to whole plant foods nutrition.

FOOD IS MEDICINE

In 2019, a true cost accounting analysis of the U.S. food system revealed that while Americans spent over $1 trillion dollars on food annually, the associated externalities impacting the natural environment and human health was calculated at $3.2 trillion dollars (The Rockefeller Foundation, 2021). Over $604 billion dollars account for the healthcare costs associated with the burden of diet-related diseases (The Rockefeller Foundation, 2021). With regard to the natural environment and biodiversity, negative externalities totaled $350 billion and $455 billion dollars, respectively (The Rockefeller Foundation, 2021). Importantly, communities of color bear an enormous burden of disease due to food insecurity, economic deprivation, structural racism, and exposure to environmental hazards (Rockefeller Foundation, 2021). This analysis illuminated the importance of restructuring the food system as an "infrastructure investment" for planetary health (Rockefeller Foundation, 2021).

"Food Is Medicine" (also known as "Food As Medicine") is a paradigm centered around creating equity, access to, and increasing the consumption of healthy foods so to combat diet-related morbidity, mortality, and food insecurity (Bleich et al., 2023). Food Is Medicine encompass an array of approaches, policies, and services. These initiatives may be in the form of medically tailored meal delivery, produce prescription programs, hands-on cooking classes, teaching kitchens, and other targeted educational and policy interventions to improve health outcomes with food (Bleich et al., 2023). Interlocking these broadscale nutrition interventions within healthcare delivery and health policy infrastructure can improve health outcomes, decrease healthcare costs, and increase equity (Mozafferian et al., 2024). Mozafferian et al. (2024) describe the importance of moving away from simply consuming enough calories known as food security to *nutrition security*—the habitual intake of nutritious foods that decreases the burden of disease and facilitates health equity.

Nutrition Prescription Programs

Nutrition Prescription Programs offer an increased intake of nutrients through whole plant foods that contain an array of phytonutrients that impart therapeutic value (Wetherill et al., 2019). Produce prescription programs are largely centered around offering free or subsidized fresh whole produce to people that are food insecure and have one or more diet-related chronic diseases (Wang et al., 2023). An enhanced meta-analysis (2013–2018) revealed that spending $42.00/month per participant in a national nutrition prescription program for 6.5 million food insecure persons in the U.S. with diabetes could result in decreased cardiovascular events and improved health outcomes while offering an overall cost savings (Wang et al., 2023).

Spotlight: Culinary Medicine at the University of Oklahoma School of Medicine

With the motto "Food is Prevention, Food Is Medicine, Food is Community", the University of Oklahoma (OU) School of Community Medicine's Culinary Medicine Program strives to reverse injustices, create food security, and increase nutrition knowledge by promoting whole plant foods nutrition and healthy lifestyle behaviors (University of Oklahoma School of Community Medicine, n.d.). In addition to training future healthcare providers on the benefits of whole foods plant-based diets, the program supports the OU Food First Pharmacy, which collaborates with the Food Bank of Eastern Oklahoma to provide prescription food packages to patients with complex co-morbidities with the goal of improving patient knowledge, skills and abilities; combatting food insecurity; and decreasing cardiovascular risk factors (Wetherill et al., 2018). In 2016–2017, their pilot intervention used prescription food packages that contained high potassium foods modeled after the Dietary Approaches to Stop Hypertension (DASH) Diet (legumes, beans, nuts, seeds and anti-inflammatory spices); they found significant increases in daily dietary fiber, fruit and vegetable intake—along with improved diastolic blood pressure in those participants presenting with high blood pressure (Wetherill et al., 2018). The OU Culinary Medicine Curriculum provides students and patients with an educational booklet that demonstrates how to prepare low-cost, whole plant foods; an excerpt is provided in Figure 10.1.

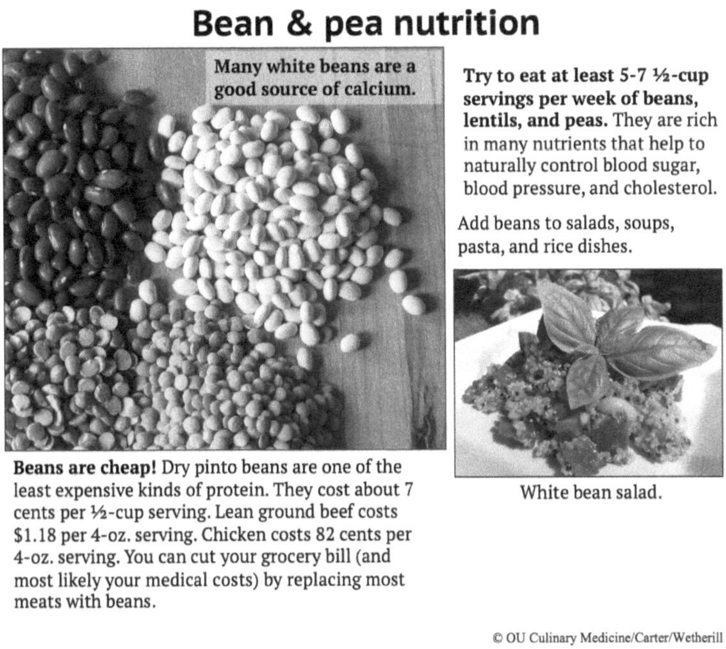

FIGURE 10.1 University of Oklahoma Culinary Medicine Book *Cooking Cures in Your Kitchen*: The Health and Economic Benefits of "Bean & Pea Nutrition", © University of Oklahoma Culinary Medicine Program, Dr. Marianna Wetherhill & Chef Valerie Carter.

Spotlight: Boston Medical Center's Food is Medicine Programs

The Preventive Food Pantry at Boston Medical Center (Boston, Massachusetts) is credited as having the first therapeutic food pantry in the nation (Boston Medical Center, n.d.). Food prescriptions are filled using groceries to ameliorate nutrition-related chronic disease where approximately 1,000,000 pounds of food is distributed a year (Boston Medical Center, n.d.). Boston Medical Center's rooftop garden supplies produce for: The Preventive Food Pantry; the onsite farmer's market; cooking demonstrations—and other educational venues (L. Hiralal, personal communication, November, 2019). The teaching kitchen is located in the center of the hospital cafeteria and is encased with glass where learning to cook fresh whole plant foods occurs alongside a large monitor that displays the rooftop garden where watching the food grow reinforces the values of this medical community (L. Hiralal, personal communication, November, 2019).

Medically Tailored Meals

Under the supervision of a Registered Dietitian Nutritionist (RDN), medically tailored meals utilize medical nutrition therapy in the form of homemade meals designed for disease amelioration for food insecure recipients that have one or more chronic diseases (Community Servings, 2019). One such program, Community Servings in Boston, Massachusetts, designs and delivers over one million meals annually (Community Servings, 2019). These meals offer one-half to two-thirds of daily nutrient needs while utilizing 16 therapeutic, culturally appropriate diets for their clients (Community Servings, 2019). A retrospective study of medically tailored meal interventions found decreases in hospitalizations, skilled nursing facility admissions, and overall medical spending amongst recipients (Berkowitz et al., 2019).

TRAUMA-INFORMED CARE AND TRAUMA-INFORMED NUTRITION

Individuals and communities that have experienced discrimination and/or neglect require sensitivities to their lived experience. Experiencing a traumatic event—whether physical or psychological—can complicate or negatively impact the choices individuals make, leading to unfavorable health outcomes (Kolis & Houston-Kolnik, 2018). Longtime exposure to extreme stressors can create a damaging buildup in the body called "toxic stress" (United Nations International Children's Emergency Fund [UNICEF], n.d.). Recognizing the impact of trauma in the delivery of health services is imperative. Trauma-informed care describes an approach for engaging with and empowering persons with a history of trauma that is centered around safety, care, and respect for the many ways that trauma may impact an individual (Kolis & Houston-Kolnik, 2018).

Trauma knows no boundaries across the life cycle (Yadav & Gunturu, 2024). Zhang and Monnat (2022) found that in the U.S., Black and Hispanic adolescents experience higher levels of adversity than their White counterparts. Indigenous Americans have experienced trauma, "...at rates higher than other U.S. population subgroups" (National Native Children's Trauma Center, 2021). Racially marginalized individuals have a higher prevalence of Adverse Childhood Experiences (ACE's), too (State of California Office of the Surgeon General, n.d.). Traumatic events in childhood may cause long-term consequences over a lifetime and may be historical or systemic, including neglect and depravation that could emerge as unhealthy behaviors including substandard food choices (California Department of Public Health & California Department of Social Services, 2020).

Trauma-informed nutrition can be characterized as, "recognizing the relationship between adversity, chronic disease, and nutritional health" (California Department of Public Health & California Department of Social Services, 2020). Negative food behaviors could manifest from a history of trauma and may include but is not limited to eating disorders, food hoarding, food addictions, and a heavy reliance on convenience foods that satisfy immediate versus long-term needs (California Department of Public Health & California Department of Social Services, 2020).

TACKLING FOOD SOVEREIGNTY AND REDEFINING FOOD SYSTEMS

Locally and globally, efforts are underway to redefine food systems that improve planetary health. In Jamaica, the Ministry of Agriculture has encouraged citizens to grow food themselves to increase access to healthy foods due to the exorbitant cost of importing food (Cave, 2013). Not only are citizens encouraged to create productive backyard gardens—but they can request a gardening kit from the government (Miles, 2024). Jamaica's National School Garden Project plans to expand and sustain schoolyard gardens while increasing parent and student engagement in agriculture, healthy behaviors, and nutrition security (Miles, 2024). In the U.S., individuals, educational institutions, and non-profit organizations continue to charge forth with re-envisioning nutrition security.

Initiatives include an array of venues including Community Supported Agriculture (CSA), productive urban farms, innovations at health centers, and inspiring individual efforts (Harmon 2019; Rangarajan & Riordan, 2019). In Detroit, Mark Covington became an urban gardening pioneer at the Georgia Street Community Garden (Beard, 2019). Covington started growing fresh produce in raised beds on 8 nearby lots, expanding to 24 lots (Beard, 2019). Detroit resident Edith Floyd founded Growing Joy Community Garden and started with 9 lots, expanding to 32 (Beard, 2019). Covington is a regular speaker about growing food at a local elementary school while Floyd hosts individuals that need to fulfill community service requirements at her urban garden (Beard, 2019). The benefits of these efforts are far-reaching as the Georgia Street Community Garden has drawn people to move to this neighborhood (Beard, 2019).

While involved in an agricultural training session at Detroit's Earthworks Urban Farm, Detroit resident Alika Muhammad said that, "Growing your own food is one of the most revolutionary acts you can participate in" (Beard, 2019). Carolyn Leadley of Detroit's Rising Pheasant Farms sees urban gardening as a way to work together, to share resources—and to transform the food system (Beard, 2019). Patrick Crouch, a former educator at Earthworks Urban Farm sees growing food as the capacity to take care of oneself without outside influences (Beard, 2019).

Gardening itself is attributed to a multitude of co-benefits. A community-based participatory research study found that Detroit's urban gardeners (n=26) positively described the physical attributes of gardening as "good exercise" and "therapeutic exercise" and noticed improved mood and stress control—along with personal growth (Beavers et al., 2022).

There are approximately 1,400 community gardens and farms in Detroit (Adams, 2019). Newell et al. (2022) studied Detroit's burgeoning urban farming movement from 2010–2016, focusing on Detroit's lower eastside. There were 53 gardens in 2010 which increased to 89 gardens in 2016 (Newell et al., 2022). Despite this impressive gain, less than 1% of vacant land in this area is used for gardening, pointing to a tremendous opportunity that is mired in the reality of the lack of home ownership and generational wealth (Newell et al., 2022).

Recipe4Health in northern California's Alameda County engages five pillars in their Food As Medicine program: safety net health centers, Medicaid health plan services, a food pharmacy, a behavioral pharmacy, and a staff training hub; they also support agriculture that is regeneratively grown by Black, Indigenous, and People of Color (Alameda County [California], 2023).

Owned by the city of Seattle and run by Seattle Tilth, the Rainier Beach Urban Farm & Wetlands boasts a 10-acre organic farm and farmstand, a CSA program, nutrition education workshops, farm walks, community dinners, cooking classes, and sliding-scale fresh produce bags (Tilth Alliance, 2025). Training courses offered to the community include "Master Composters/ Sustainability Stewards"; "Soil & Water Stewards"; "Youth Stewards"; and the youth "Go Green! Garden Club" (Tilth Alliance, 2025).

Food for Lane County (Eugene, Oregon) is a non-profit food bank with an array of programs designed to create access to healthy foods for individuals and families that are marginalized while offering training programs designed to create equity (Food For Lane County, 2025). Programs include: a paid training program where graduates are certified in food warehouse skills; a paid working farm training program for low-income youth; a 2.5 acre in-city farm complete with a compost demonstration site and outdoor kitchen; free gardening classes that include seeds and educational materials; and free nutritious meals or food boxes (Food for Lane County, 2025). This diverse array of programming serves a community where nearly a quarter of all children are impoverished and over half of school-aged kids qualify for the free national school lunch program (Food for Lane County, 2018).

Despite the impressive range of individuals and programs described in this chapter, the need for widespread structural change is evident. Until policy makers create systematic change, individuals, organizations, and communities must continuously strive to create nutrition security and equity in the most constructive ways possible.

NOTE

1 Jones, J. R. (2020, April 30). *Culinary cure.* Oklahoma University School of Community Medicine. www.tulsapeople.com/lifestyle/culinary-cure/article_e5480be6-8b0e-11ea-a1ee-037bde2848c3.html

REFERENCES

Adams, B. (2019, November 5). In Detroit, a new type of agricultural neighborhood has emerged. *YES!* www.yesmagazine.org/social-justice/2019/11/05/food-community-detroit-garden-agriculture

Alameda County [California]. (2023). *Join the food as medicine movement.* https://recipe4health.acgov.org/

Beard, R. (2019, October, 20). *Motown to growtown: Detroit's urban farming revolution.* Rewind. *Al Jazeera.* www.youtube.com/watch?v=I_7rzoFzV08

Beavers, A.W., Atkinson, A., Varvatos, L.M., Connolly, M. & Alaimo,K. (2022). How gardening in Detroit influences physical and mental health. *International Journal of Environmental Research in Public Health 19*(7899), 1–11. https://doi.org/10.3390/ijerph19137899

Berkowitz, S.A., Terranova, J., Randall, L., Cranston, K., Waters, D.B., & Hsu, J. (2019). Association between receipt of a medically tailored meal program and health care use. *JAMA Internal Medicine*, *179*(6), 786–793. https://doi.org/10.1001/jamainternmed.2019.0198

Bleich, S.N., Dupuis, R., & Seligman, K. (2023). Food Is Medicine movement-key actions inside and outside the government. *JAMA Health Forum*, *4*(8), e233149. https://doi.org/10.1001/jamahealthforum.2023.3149

Boston Medical Center. (n.d.). *Preventive food pantry.* www.bmc.org/nourishing-our-community/preventive-food-pantry

California Department of Public Health & California Department of Social Services. (2020). *Trauma-informed nutrition.* Essentials for Childhood Initiative.

Cave, D. (2013, August 3). As cost of importing food soars, Jamaica turns to the Earth. *The New York Times.* www.nytimes.com/2013/08/04/world/americas/as-cost-of-importing-food-soars-jamaica-turns-to-the-earth.html

Community Servings. (2019*). Medically tailored meals & healthcare utilization.* www.servings.org/wp-content/uploads/2019/04/RWJF_Community-Servings-White-Paper-2019_SP_web.pdf

Food For Lane County. (2025). *Programs & services.* www.foodforlanecounty.org/go-learn-more/programs-services/

Food For Lane County. (2018). *Hunger in Lane County.* www.foodforlanecounty.org/wp-content/uploads/2019/03/HFS-2019-FINAL.pdf

Harmon, A.H. (2019). Taking it outside: The value added by experiential learning on food systems. In Hawkins, I.W. (Ed.), *Promoting biodiversity in food systems* (pp. 231–245). CRC Press.

Kolis, K., & Houston-Kolnik, J. (2018). *Trauma types and promising approaches to assist survivors.* Illinois Criminal Justice Authority. https://icjia.illinois.gov/researchhub/articles/trauma-types-and-promising-approaches-to-assist-survivors

Miles, K. (2024, May 22). School garden initiative 'planting' agriculture into curriculum. *Jamaica Observer.* www.jamaicaobserver.com/2024/05/22/school-garden-initiative-planting-agriculture-curriculum/#:~:text=FOLLOWING%20its%20launch%20earlier%20this,their%20practical%20experience%20in%20agriculture

Mozafferian, D., Aspry, K.E., Garfield, K., Kris-Etherton, P., Seligman, H., Velarde, G.P., Williams, K., & Yang, E. (2024). "Food is medicine" strategies for nutrition security and cardiometabolic health equity. *Journal of the American College of Cardiology*, *83*(8), 843–864. https://doi.org/10.1016/j.jacc.2023.12.023

National Native Children's Trauma Center. (2021). *Fact sheet: Trauma exposure among American Indians and Alaska Natives.* https://static1.squarespace.com/static/5859aae1ff7c503bc4a5a155/t/6089ab127add03492ac1a345/1619634963501/NNCTC+Fact+Sheet+Trauma.pdf

Newell, J.P., Foster, A., Borgman, M., & Meerow, S. (2022). Ecosystem services of urban agriculture and prospects for scaling up production: A study of Detroit. *Cities*, *125*, 1–14. https://doi.org/10.1016/j.cities.2022.103664

Rangarajan, A., & Riordan, M. (2019). *The promise of urban agriculture: National study of commercial farming in urban areas.* United States Department of Agriculture/Agricultural Marketing Service and Cornell University Small Farms Program.

State of California Office of the Surgeon General. (n.d.). *ACEs and toxic stress: A public health crisis.* https://osg.ca.gov/wp-content/uploads/sites/266/2020/12/Overview-of-ACEs-and-Toxic-Stress_CA_OSG-Report_12092020.pdf

The Rockefeller Foundation. (2021, July). *Measuring what matters to transform the U.S. food system.* www.rockefellerfoundation.org/wp-content/uploads/2021/07/True-Cost-of-Food-Full-Report-Final.pdf

Tilth Alliance. (2025). *Rainier Beach urban farm & wetlands.* https://tilthalliance.org/our-work/urban-farms-gardens/rainier-beach-urban-farm-wetlands/

United Nations International Children's Emergency Fund [UNICEF] (n.d.). *What is toxic stress?* www.unicef.org/parenting/mental-health/what-is-toxic-stress#toxic-stress

University of Oklahoma School of Community Medicine. (n.d.). Culinary medicine program . www.ou.edu/tulsa/community_medicine/about-scm/culinary-medicine

Wang, L., Lauren, B.N., Hager, K., Zhang, F.F., Wong, J.B., Kim, D.D., & Mozaffarian, D., (2023). Health and economic impacts of implementing produce prescription programs for diabetes in the United States: A microsimulation study. *Journal of the American Heart Association, 12*(e029215), 1–21. https://doi.org/10.1161/JAHA.122.029215_

Wetherhill, M.S., Whelan, L.J., Upton, L., Schumann, S., Davis, G.C., Carter, V.P., & Tollefson, M. (2019). Nutrition prescriptions made simple. *The Journal of the Oklahoma State Medical Association, 112*(10), 370–377.

Wetherhill, M.S., McIntosh, H.C., Beachy, C., & Shadid, O. (2018). Design and implementation of a clinic-based food pharmacy for food insecure, uninsured patients to support chronic disease self-management. *Journal of Nutrition Education and Behavior, 50*(9), 947–949. https://doi.org/10.1016/j.jneb.2018.05.014

Yadav G., & Gunturu, S. (2024). Trauma-informed therapy. National Institutes of Health, StatPearls. www.ncbi.nlm.nih.gov/books/NBK604200/

Zhang, X., & Monnat, S.M. (2021). Racial/ethnic differences in clusters of adverse childhood experiences and associations with adolescent mental health. *SSM-Population Health, 17*, 100997. https://doi.org/10.1016/j.ssmph.2021.100997

Bolstering Nature for Planetary Health

11

"You cannot protect the environment unless you empower people, you inform them, and you help them understand that these resources are their own, that they must protect them."

Dr. Wangari Maathai[1], Nobel Laureate, Parliamentarian, Founder of the Greenbelt Movement.

OVERVIEW

This chapter focuses on bolstering nature to improve planetary health and mitigate environmental injustices. As the remaining fauna and flora have limited places to survive amidst record-setting heat and climate extremes—every effort to restore and connect with nature on all levels is vital. When restoring nature, we create a multitude of co-benefits and can forge deep bonds. Importantly, when youth are supported in their concern or connection to nature, they may feel emboldened as adults to undertake pro-environmental leadership roles (Hawkins et al., 2015). This chapter will start with our declared fundamental right to a safe and healthy natural environment followed by an array of actions and opportunities to bolster and relish nature by combatting environmental injustices; creating biologically dynamic and resilient ecosystems; contributing to community science—and preventing light pollution.

DOI: 10.1201/9781003153528-13

OUR HUMAN ENVIRONMENTAL RIGHTS AND THE ENVIRONMENTAL RULE OF LAW

Although not yet enforceable, the United Nations General Assembly (2022) voted to declare our fundamental human right to a clean, healthy, and safe natural environment while stressing that the global community must uphold these principles (United Nations General Assembly, 2022). The Environmental Rule of Law demands that natural resources are protected to support healthy ecosystems (United Nations Environment Program, 2019). However, the first global report on the Rule of Environmental Law found a lack of political will and ineffective governance around the globe including funding disparities, backlashes, sanctions that prevent protection of the natural environment—and extreme violence where some defenders of the natural environment were murdered (United Nations Environment Program, 2019). The goal of the Environmental Rule of Law framework is upholding environmental laws, creating accountability—and continuous evaluation (United Nations Environment Program, 2019). Enforcing the Environmental Rule of Law can result in higher per-capita incomes in resource-extractive countries and improved well-being which coincides with the Sustainable Development Goals noted in Chapter 1 (United Nations Environment Program, 2019).

NATURE-POSITIVE ACTIONS

The stability of the Earth System will depend upon ensuring all remaining natural ecosystems remain intact while preventing further damage—and simultaneously bolstering biodiversity (Locke et al., 2021). That is, as a global society, we retain and restore species, ecosystems, and the Earth's natural processes on all levels—from local landscapes to country-wide restoration and ultimately, global transformation (Nature Positive Initiative, 2023; Soliveres et al., 2016). The metrics-driven Nature Positive Initiative is a worldwide forum "halting and reversing biodiversity loss by 2030…to set the path for full recovery of nature by 2050" as depicted in Figure 11.1 (The Nature Positive Forum, n.d.). The accompanying Nature Positive Forum promotes action and synergies to accomplish this goal with a "whole-of-society" approach (The Nature Positive Forum, n.d.).

The fifteenth Conference of the Parties (COP-15) adopted the 2023 Kunming-Montreal Global Biodiversity Framework's "Biodiversity Plan

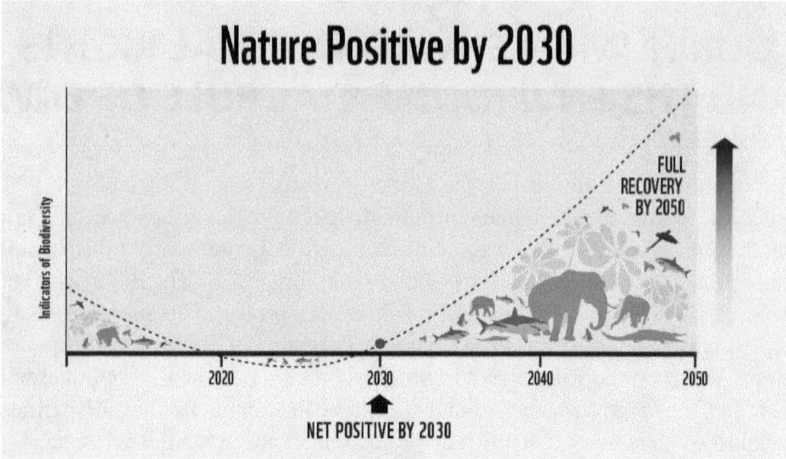

FIGURE 11.1 The Nature Positive Forum "Nature Positive by 2030".

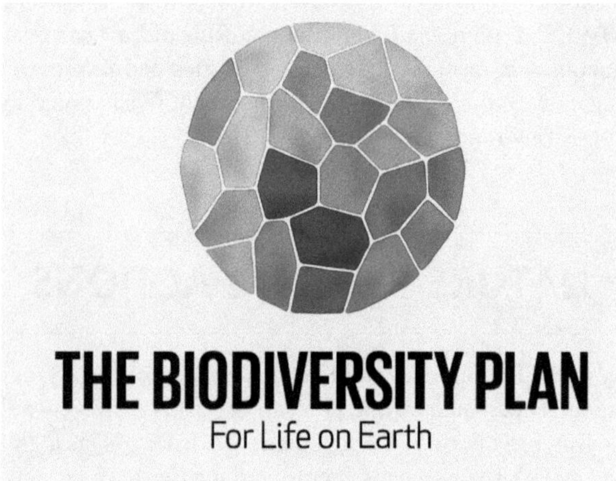

FIGURE 11.2 The Kunming-Montreal Global Biodiversity Framework Logo, 2023.

2030" that contains 23 target areas (Convention of Biological Diversity, 2024). The first phase of their public campaign is entitled, "The Biodiversity Plan: For Life on Earth" whose logo is displayed in Figure 11.2:

Target 12 of the 2023 Kunming-Montreal Global Biodiversity Framework aims to "Enhance Green Spaces and Urban Planning for Human Well-Being and Biodiversity". One such effort comes from challenging city mayors across the globe to plant trees to improve planetary health (United Nations

Economic Commission for Europe, n.d.a.). Since 2019, over 80 cities across the globe including Aarhus (Denmark), Ashgabat (Turkmenistan), Denver (Colorado, USA), Mexico City (Mexico) and Tirana (Albania) have planted nearly 11 million trees as of December 31, 2023 (United Nations Economic Commission for Europe, n.d.b.).

REWILDING, NATURESCAPING AND MICRO FORESTS

On the ground and across the globe, citizens are utilizing nature to rectify injustices, create equity, improve planetary health—and to educate and empower. Ghana (West Africa) offers an important example. Ghana lost nearly half of its forests in recent years because of cocoa production, gold mining, and illegal harvesting (Biggar, 2019). To combat this destruction, 41 million trees have been planted with support from government interventions, youth-led initiatives—and private and public organizations (Appiah-Dolphyne, 2024).

Rewilding is a name given to restoring ecosystems as they functioned before human extirpation that's described as giving fauna, flora, and the land back to each other (Davis, 2018). When trophic levels of ecosystems are abundant—the Earth System becomes more resilient (Ripple et al., 2022; Soliveres et al., 2016). Rewilding enhances trophic complexity where in the Swiss National Park (Switzerland), a strict non-human management approach has resulted in the recovery of rare or nearly extinct species (Perino et al., 2019).

Policy is critically important to rewilding efforts. Oregon's House Bill 3464 offers protection to beavers and recognizes their value in the multifaceted climate crises (Oregon State Legislature, 2023). Beavers are a keystone species where their role in the integrity of entire ecosystems is invaluable; they've been described as "ingenious creatures", "ecosystem engineers" and "the ultimate keystone species" (U.S. National Park Service, 2021; Newton, 2021). Figure 11.3 denotes some of the distinguished features of the North American Beaver, *Castor canadensis* (U.S. Forest Service, n.d.).

Scientists have determined that restoring beaver populations is vital to rewilding strategies (Ripple et al., 2022). Whether it's California, Britain, or Germany—reintroducing beavers creates a multitude of co-benefits including wetland formation (that minimizes the impact of floods); establishing wildlife havens; and increasing carbon capture (Begum, 2021; California Department of Fish and Wildlife, 2025; Newton, 2021).

Naturescaping can be described as a way for humans to enter into partnership with the land by recreating Indigenous ecosystems (Oregon Department

FIGURE 11.3 Distinguished Physiological Characteristics of the North American Beaver, U.S. Forest Service.

of Fish and Wildlife, 2015). This includes attracting and offering biodiversity healthy places which includes but is not limited to planting native plants of varied colors; offering nest boxes and a source of water for hydration and bathing; allowing leaves to accumulate; eliminating the use of agrichemicals; offering rock piles; and recreating the respective ecoregion's forest by offering multiple layers of habitat from canopy trees down to soil fungi (Oregon Department of Fish and Wildlife, 2015).

Micro forests (also known as pocket forests or mini forests or tiny forests) can quickly offer refuge for biodiversity (LA Micro Forests, 2022; Miyawaki, 1999). This concept is derived from the approach of the Japanese conservationist Akira Miyawaki where multilayered, dense plantings facilitate rapid growth, attracting more biodiversity than traditional reforestation efforts (LA Forests; Miyawaki, 1999). Miyawaki's method of green space rejuvenation using "native forests by native trees" resembles Indigenous ecosystems where the dense, mixed plantings can manage independent of human intervention (Miyawaki, 1999).

A variety of tools and resources are now available for those without expertise in naturescaping. Municipalities, land-grant university programs and organizations such as the California Native Plant Society (calscape.org) offer native plant databases and landscape design tools (California Native Plant Society, 2025). The Pennsylvania Department of Conservation and Natural Resources offers "Native Garden Templates" for a variety of settings as well as guidance on attracting wildlife (Pennsylvania Department of Conservation and Natural Resources, 2024). The Minnesota Board of Water and Soil Resources "Lawn to Legumes" program partners with organizations such as Blue Thumbs and Metro Blooms to enable residents to plant native plants by offering a myriad of resources such as money, free trees and mulch, workshops, coaching—while also prioritizing equity (Minnesota Board of Water and Soil Resources, n.d.; Metro Blooms, n.d.).

Los Angeles based native plant horticulturist Katherine Pakradouni utilizes Miyawaki's micro forest methods in environmental restoration efforts while also offering a free, comprehensive planning tool for the public online at lamicroforest.com (Seeds to Landscape, 2022). Pakradouni's "plant calculator" is specific to California ecoregions where the entire process from planning to aftercare is delineated (Seeds to Landscape, 2022). Pakradouni emphasizes, "…there is no space left anymore that should be just for people" (Marantos, 2022).

The Washington State Department of Fish & Wildlife offers a "Habitat at Home" program (also known as Wildlife Habitat Yards) where individuals can certify their outdoor space after meeting the requirements of offering food, water, and shelter for wildlife (Washington State Department of Fish & Wildlife, 2024a). This program emphasizes conservation measures such

FIGURE 11.4 Washington State Department of Fish and Wildlife Certified Backyard Wildlife Sanctuary Signage and Associated Biodiversity, Seattle, Washington. Photos © Irana Hawkins.

as eliminating the use of agrichemicals; keeping cats indoors; using mulch to protect soils; and using compost to bolster soil fertility (Washington State Department of Fish & Wildlife, 2024a). Upon certification, individuals can display an educational recognition sign as seen in Figure 11.4 where the signs

FIGURE 11.4 (Continued)

to the left convey the importance of a wildlife snag while the deliverables are emphasized in the photo on the right where the snag's woodpecker holes represent the food source it offers along with creating habitat for the wild Eastern Cottontail rabbit (*Sylvilagus floridanus*).

NATIVE PLANTS AND REWILDING

As mentioned in Chapter 8, native plants have evolved to the singular circumstances of their respective ecoregion long before human perturbances. Non-native plants have led to decreases in bird populations and reproductive distress while some non-native plants are not well-suited for Indigenous biodiversity across trophic levels (Mata et al., 2021; Narango et al., 2018). Assessing a plant-insect meta-network across urban greenspaces in Melbourne, Australia, Mata et al. (2021) found that Indigenous plants better supported the highest numbers of Indigenous insect biodiversity. Greenspaces with multi-layered vegetation provide an important array of resources for a diverse range of biodiversity (Mata et al., 2021). Altogether, native biodiversity is indispensable for high-performing, multifunctioning ecosystems (Lefcheck et al., 2015; Soliveres et al., 2016).

Construction is underway at the Wallis Annenberg Wildlife Crossing in southern California, the largest wildlife bridge in the world to date (State of California, 2024). Importantly, this bridge will not only offer critical habitat for wildlife and a desperately needed passageway—but will utilize "hyperlocal" native plants grown from seed (Marantos, 2023). It was the plight of a local puma named "P-22" in the Los Angeles area and the threat of extinction of the pumas of the Santa Monica mountains that helped galvanize support for the wildlife bridge (Nelson, 2023; National Park Service, 2023). P-22 was euthanized in 2023 due to severe injuries including the sequelae from a car strike; thousands of people attended P-22's celebration of life service, demonstrating our connection to and concern for wildlife (Nelson, 2023).

NATURE AND HEALTH EQUITY

A special report of the Trust For Public Land (n.d.) found greenspace disparities in communities of color where parks are 50% smaller and serve five times the number of people compared to White communities. Temperatures are lower in communities with trees and greenspace, mitigating the urban heat effect (Trust For Public Land, n.d.). Overall, approximately one in three Americans don't have a park within a 10-minute walk from their residence (Trust for Public Land, n.d.).

Amongst the most populous cities in the U.S., those residing in cities with more parks are healthier than those with fewer parks as evidenced by improved mental health outcomes and increased levels of physical activity (Foderaro &

Klein, 2023). Kim et al. (2023) assessed the relationship between long-term exposure (over 20 years) to greenspace and epigenetic age and found that Black study participants were less likely to have exposure to greenspace compared to White participants and hence, did not benefit from slower epigenetic aging. Subgroup analyses found that those with limited economic resources may benefit more from access to greenspace (Kim et al., 2023). Rigolon et al. (2021) found that greenspace contributed more to health equity for people of lower socioeconomic status than more affluent persons.

Using a health-impact assessment approach, Garber et al., (2023) projected that increasing 30% of land area with native plants in Denver (Colorado) could prevent premature deaths. A systematic review of longitudinal studies assessing greenspace and all-cause mortality in seven countries found a significant relationship between decreased all-cause mortality and greenspace within 500 meters of one's residence (Rojas-Rueda et al., 2019). Another study (2003–2015) comparing individual healthcare costs and greenspace of over 5 million health insurance members in northern California found a significant association between higher residential greenspace and lower direct healthcare costs (Van Den Eeden et al., 2022).

With regard to crime, reviews of the literature generally find that green space is associated with decreased crime (Shepley et al., 2019). Jay et al. (2022) analyzed census tracts for racialized economic segregation, tree canopy, and gun violence in six U.S. cities and found that greater financial privilege was associated with more tree canopy and less gun violence (Jay et al., 2022). Sacramento (California) is investing in the urban tree canopy in underserved neighborhoods to rectify injustices, decrease urban heat, mitigate climate change—and to increase livability (Matsui, 2024). As trees increase property values and economic benefits for homeowners, they can help facilitate generational wealth (Thompson et al., 2021).

Nature prescription programs are specific recommendations by healthcare providers to spend time with nature (Nguyen et al., 2023). Nguyen et al. (2023) found that persons engaged in nature prescription programs increased their daily step count, experienced improved blood pressure—and had improved anxiety and depression scores compared to the control group, respectively (Nguyen et al., 2023).

Accessing Nature

Josè G. Gonzâlez, the founder of Latino Outdoors whose mission is to engage Latino communities in the outdoors believes, "Nature is essential. We really need to see it as a necessity rather than just a nicety" (Sarah, 2020). As access to greenspace in underserved communities is limited, organizations have offered important opportunities to access nature. In Los Angeles (California), the Skyhook Foundation partners with the Los Angeles Unified School District to bring science, technology, engineering, art, and math opportunities to diverse, underserved children (Skyhook Foundation, 2025). The program "Camp Skyhook" offers children a weeklong overnight experience in the Angeles National Forest (Skyhook Foundation, 2025). The program has a six-year waitlist, a testament of its value (Skyhook Foundation, 2025).

Despite close proximity to the Pacific Ocean, underserved students participating in "Camp Cove" at the Crystal Cove Conservancy in Laguna Beach (California) may not have access to the beach—or have never been to the beach (Crystal Cove Conservancy, 2025). Through a unique multi-agency partnership, close to 100 students have participated in a weeklong, daytime immersion event, learning about ocean ecology and natural history—even participating in the release of a rehabilitated elephant seal (Crystal Cove Conservancy, 2025). The experience has been described as transformative for these youngsters (Crystal Cove Conservancy, 2025).

THE VALUE OF COMMUNITY SCIENCE

The California Academy of Sciences describes community science as, "a global movement through which scientists and non-scientists alike make observations, collect data, and help answer some of our planet's most pressing questions" (California Academy of Sciences, 2024). Because of the dire circumstances of biodiversity and the natural environment, community science is an invaluable tool where non-governmental and governmental agencies alike stress the importance of community science—and encourages everyone to participate.

As an example, the International Monarch Monitoring Bioblitz July 26 to August 4, 2024 relied on community scientists across three nations to document the entire lifecycle of the Monarch butterfly along with their native milkweed host plants (Koenig, 2024). A "bioblitz" is a targeted effort to document as many observations of a certain species—or a variety of species—within a unique ecosystem and timeframe; thereafter, trained scientists assess the data that can inform future actions.

Internet-based tools such as iNaturalist (iNaturalist.org) whose mission it is to, "...connect people to nature and advance biodiversity science and conservation" offers engagement through photo or video observation (iNaturalist, 2024). Once corroborated, these observations are shared with the Global Biodiversity Information Facility, an intergovernmental, open-access global data platform (California Academy of Sciences, n.d.). Agencies such as the Washington State Department of Fish & Wildlife encourages community science using iNaturalist (iNaturalist.org), eBird (ebird.org), Carnivore Spotter (carnivorespotter.org), or Bat Activity Trends (bat.fieldscope.org) (Washington Department of Fish and Wildlife, 2024b).

THE IMPORTANCE OF DARK SKIES

As mentioned in Chapter 2, artificial light at night (ALAN) can negatively impact the biology of animals and plants, altering ecosystem services, reproduction, and even survival (DarkSky International, 2025). Natural magnetic sensing capabilities of migratory birds may be impaired while nocturnal pollinators are also disrupted (DarkSky International, 2017; DarkSky International, 2019). The services of nocturnal pollinators are far-reaching and includes pollinating commercially important crops and endemic plants (Robertson et al., 2021). Bats pollinate over 300 species of fruit while nocturnally pollinating flies, moths, and net-winged insects offer significant contributions to apple production while also complimenting daytime pollination services (Robertson et al., 2021; U.S. Forest Service, n.d.).

More specifically, 83% of the global population endure light-polluted skies while 60% of Europeans and 80% of North Americans cannot see the Milky Way (Falchi et al., 2016). Finally, although using outdoor lighting at

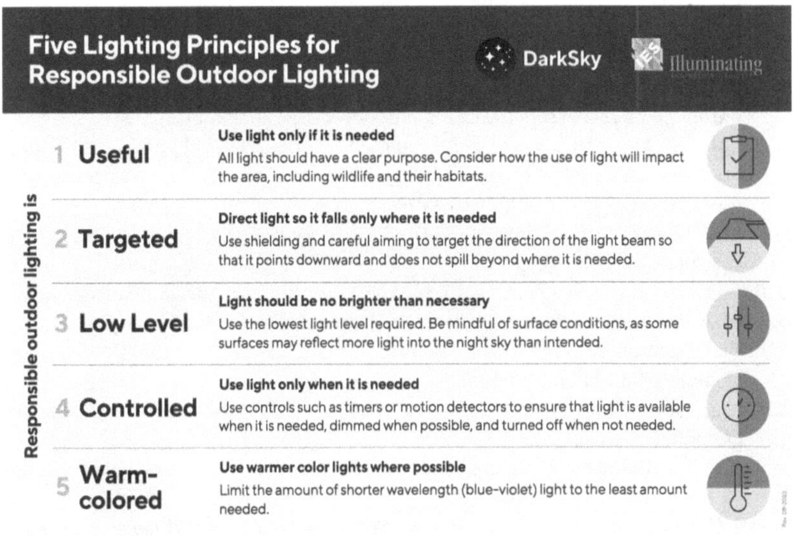

FIGURE 11.5 Five Lighting Principles for Responsible Outdoor Lighting, Dark Sky International & Illuminating Engineering Society, 2024.

night may be associated with the perception of increased safety, this has not been corroborated in the literature (DarkSky International, 2023). As lights can be used with integrity at night, this chapter will conclude with the DarkSky International & Illuminating Engineering Society's five principles for outdoor lighting, as noted in Figure 11.5.

NOTE

1 Dixon, R. (2011, September 27). Kenyan environmentalist won Nobel. *The Los Angeles Times.* www.latimes.com/archives/la-xpm-2011-sep-27-la-me-wangari-maathai-20110927-story.html

REFERENCES

Appiah-Dolphyne, J. (2024, June 5). *Green Ghana Day: About 30 million seedlings survived out of 41 million planted since 2021, says Abu Jinapor.* Asaase Broadcasting Company; Asaase Radio. www.asaaseradio.com/green-ghana-day-about-30-million-seedlings-survived-out-of-41-million-planted-since-2021-says-abu-jinapor/

Begum, T. (2021, April 7). *Record numbers of beavers are being introduced to the UK.* Natural History Museum London. www.nhm.ac.uk/discover/news/2021/april/record-numbers-of-beavers-are-being introduced-to-the-uk.html

Biggar, H. (2019, April). *A future with 20 million new trees.* Global Landscapes Forum. https://thinklandscape.globallandscapesforum.org/34177/in-ghana-one-man-plants-a- future-with-20-million-new-trees/

California Academy of Sciences. (2024). *Community science.* www.calacademy.org/community-science

California Academy of Sciences. (n.d.). *Using iNaturalist for community science.* www.calacademy.org/community-science/using-inaturalist-for-community-science

California Department of Fish and Wildlife. (2025). *Beaver.* https://wildlife.ca.gov/Conservation/Mammals/Beaver

California Native Plant Society. (2025). *Calscape: California's native plant destination.* https://calscape.org

Convention of Biological Diversity. (2024, October 1). *The biodiversity plan: For life on Earth.* www.cbd.int/gbf

Crystal Cove Conservancy. (2025). *Camp Cove returns to Crystal Cove State Park.* https://crystalcove.org/camp-cove-returns-to-crystal-cove-state-park/

DarkSky International. (2025). *What is light pollution.* https://darksky.org/resources/what-is-light-pollution/

DarkSky International. (2023, October 15). *Outdoor lighting at night doesn't do what you think it does to reduce crime and increase safety*. https://darksky.org/resources/what- is-light-pollution/effects/safety/#:~:text=of%20light%20pollution,Outdoor%20lighting%20at%20night%20doesn't%20do%20what%20you%20think,reduce%20crime%20and%20increase%20safety&text=What%20is%20light%20pollution%3F,-Effects%20of%20light&text=There%20is%20no%20clear%20scientific,shown%20to%20make%20us%20sa fer

DarkSky International. (2019, May 9). *Light pollution poses threat to migrating birds*. https://darksky.org/news/light-pollution-poses-threat-to-migrating-birds/

DarkSky International. (2017, August 7). *Lights at night turn nocturnal pollinators off*. https://darksky.org/news/lights-at-night-turn-nocturnal-pollinators-off/

Davis, J. (2018, July 21). *Rewilding, distilled*. Rewilding Earth. https://rewilding.org/rewilding-distilled/

Falchi, F., Cinzano, P., Duriscoe, D., Kyba, C.C.M., Elvidge, C.D., Baugh, K., Portnov, B.A., Rybnikova, N.A. & Furgoni, R. (2016). The new world atlas of artificial night sky brightness. *Science Advances*, *2* (e1600377), 1–25.

Foderaro, L.W., & Klein, W. (2023, May 23). *The power of parks to promote health*. Trust for Public Land. www.tpl.org/parks-promote-health-report

Garber, M.D., Guidi, M., Bousselot, J., Benmarhnia, T., Dean, D., & Rojas-Rueda, D. (2023). Impact of native-plants policy scenarios on premature mortality in Denver: A quantitative health impact assessment. *Environment International*, *178*, 1–9.

Hawkins, I.W., Balsam, A.L., & Graves, D. (2015). A qualitative study of how Registered Dietitians made the connection between diet, climate change, and environmental degradation. *Journal of Hunger & Environmental Nutrition*, *10*(1), 47–59. https://doi.org/10.1080/19320248.2015.1004213

iNaturalist. (2024, September 10). *About iNaturalist*. www.inaturalist.org/pages/about

Jay, J., Kondo, M.C., Lyons, V.H., Gause, E., & South, E.C. (2022). Neighborhood segregation, tree cover and firearm violence in 6 U.S. cities, 2015–2020. *Preventive Medicine*, *165*, 1–7. https://doi.org/10.1016/j.ypmed.2022.107256

Kim, K., Joyce, B.T., Nannini, D.R., Zheng, Y., Gordon-Larsen, P., Shikany, J.M., ... & Hou, L. (2023). Inequalities in urban greenness and epigenetic aging: Different associations by race and neighborhood socioeconomic status. *Science Advances*, *9*, 1–10.

Koenig, M. (2024, July 22). *Help raise awareness and support Monarch butterfly conservation: Join the 2024 Monarch Blitz, July 26 – August 4!* [Press release]. U.S. Department of Fish and Wildlife. www.fws.gov/story/2024-07/join-monarch-blitz-july-26-august-4

LA Micro Forests. (2022). *Background: Micro forests*. Seed to Landscape. www.lamicroforests.com/background

Lefcheck, J.S., Byrnes, J.E., Isbell, F., Gamfeldt, L., Griffin, J.N., Eisenhauer, N., ... & Duffy, J.E. (2015). Biodiversity enhances ecosystem multifunctionality across trophic levels and habitats. *Nature Communications*, *6*(6936), 1–7. https://doi.org/10.1038/ncomms7936

Locke, H., Rockström, J., Bakker, P., Bapna, M., Gough, M., Hilty, J., Lambertini, M., Morris, J., Rodriguez, C.M., Samper, C., Sanjayan, M., Zabey, E., & Zurita, P., (2021, April 30). *A nature-positive world: The global goal for nature*. Post-2020

Pavilion Partnership. www.nature.org/content/dam/tnc/nature/en/documents/NaturePositive_GlobalGoalCEO.pdf

Mata, L., Andersen, A.N., Morán-Ordóñez, A., Hahs, A.K., Backstrom, A., Ives, C.D., ... & Lynch, Y. (2021). Indigenous plants promote insect biodiversity in urban greenspaces. *Ecological Applications*, *31*(4), e02309. https://doi.org/10.1002/eap.2309

Marantos, J. (2022, August 3). Why L.A. needs more micro forests (and how to plant your own). *Los Angeles Times.* www.latimes.com/lifestyle/story/2022-08-03/how-to-plant-microforests-in-los-angeles

Marantos, J. (2023, June 21). Saving the next P-22 starts with a million 'hyperlocal' seeds and a bare-bones nursery. *Los Angeles Times.* www.latimes.com/lifestyle/story/2023-06-21/wallis-annenberg-wildlife-crossing-agoura-hills-hyperlocal-seeds-native-plants

Matsui, D. (2024, April 25). *City unveils their draft Sacramento urban forest plan.* [Press release]. https://matsui.house.gov/media/press-releases/matsui-and-city-sacramento-annnounce-nearly-2-million-federal-funding-citys

Metro Blooms. (n.d.). *Low-cost and accessible gardening resources.* https://metroblooms.org/low-cost-and-accessible-gardening-resources/

Minnesota Board of Water and Soil Resources. (n.d.). *Lawns to legumes: Your yard can BEE the change.* https://bwsr.state.mn.us/l21

Miyawaki, A. (1999). Creative ecology: Restoration of native forests by native trees. *Plant Biotechnology*, *16*(1), 15–25.

Mi Narango, D.L., Douglas, W. Tallamy, D.W., & Marra, P.P. (2018). Nonnative plants reduce population growth of an insectivorous bird. *Proceedings of the National Academy of Sciences*, *115*(45), 11549–11554. https://doi.org/10.1073/pnas.1809259115

National Park Service. (2023, June 13). *Puma profiles: P-22.* www.nps.gov/articles/000/puma-profiles-p-22.htm#:~:text=P%2D22%20was%20our%20most,the%20city%20of%20Los%20Angeles.

Nature Positive Forum. (n.d.). *The nature positive forum.* www.naturepositive.org/app/uploads/2024/02/Nature-Positive-Forum-description.pdf

Nature Positive Initiative. (2023, November 27). *The definition of nature positive.* https://4783129.fs1.hubspotusercontentna1.net/hubfs/4783129/The%20Definition%20of%20Nature%20Positive.pdf

Nelson, L.J. (2023, February 5). Memorial for mountain lion P-22, 'the king of Griffith Park,' draws thousands. *Los Angeles Times.* www.latimes.com/california/story/2023-02-04/thousands-gather-to-attend-p-22-celebration-of-life-at-greek-theater

Newton, E. (2021). *Bavarian beaver adventure: A tale of coexistence and conservation.* Citizen Zoo. www.citizenzoo.org/bavarian-beaver-adventure-a-tale-of-coexistence-and-conservation/

Nguyen, P., Astell-Burt, T., Rahimi-Ardibili, H., & Fong, X. (2023). Effect of nature prescriptions on cardiometabolic and mental health, and physical activity: A systematic review. *Lancet Planetary Health*, *7*, e313–e328.

Oregon Department of Fish and Wildlife. (2015, July 23). *Naturescaping: A landscaping partnership with nature.* www.dfw.state.or.us/conservationstrategy/naturescaping/index.asp

Oregon State Legislature. (2023). *House Bill 3464*. 82nd Oregon Legislative Assembly, Regular Session. https://olis.oregonlegislature.gov/liz/2023R1/Measures/Overview/HB3464

Pennsylvania Department of Conservation and Natural Resources. (2024). *Native garden templates*. www.dcnr.pa.gov/Conservation/WildPlants/LandscapingwithNativePlants/NativeGardenTemplates/Pages/default.aspx

Perino, A., Pereira, H.M., Navarro, L.M., Fernandez, N., Bullock, J.M., Ceaus, S., Cortes-Avizanda, van Klink, R., Kuemmerle, T., Lomba, A., Pe'er, G., Plieninger, T., Rey, J.M., Benayas, J.M.R., Sandom, C.J., Svenning, J., & Wheeler, H.C. (2019). Rewilding complex ecosystems. *Science*, *364*(351), 1–8.

Rigolon, A., Browning, M.H.E.M., McAnirlin, O., & Yoon, H. (2021). Green space and health equity: A systematic review on the potential of green space to reduce health disparities. *International Journal of Environmental Research and Public Health*, *18*(2563), 1–27. https://doi.org/10.3390/ijerph18052563

Ripple, W.J., Wolf, C., Phillips, M.K., Beschta, R.B., Vuceitich, J.A., Kauffman, J.B., Law, B.E., Wirsing, A.J., Lambert, J.E., Leslie, E., Vynne, C., Dinerstein, E., Noss, R., Wuerthner, G., Dellasala, D.A., Bruskotter, J.T., Nelson, M.P., Christ, E., Darimont, C., & Ashe, D.M. (2022). Rewilding the American west. *BioScience*, *72*(10), 931–935.

Robertson, S.M., Dowling, A.P.G., Wiedenmann, R.N., Joshi, N.K., & Westerman, E.L. (2021). Nocturnal pollinators significantly contribute to apple production. *Journal of Economic Entomology*, *114*(5), 2155–2161. https://doi.org/10.1093/jee/toab145

Rojas-Rueda, D., Nieuwenhuijsen, M.J., Gascon, M., Perez-Leon, D., & Mudu, P. (2019). Green spaces and mortality: A systematic review and meta-analysis of cohort studies. *Lancet Planetary Health*, *3*, e469–e477.

Sarah, L. (2020, July 30). *Nature for all: The groups bringing inclusion outdoors*. KQED. www.kqed.org/news/11830597/nature-for-all-the-groups-bringing-inclusion-outdoors

Shepley, M., Sachs, N., Sadatsafavi, H., Fournier, C., & Peditto, K. (2019). The impact of green space on violent crime in urban environments: An evidence synthesis. *International Journal of Environmental Research and Public Health*, *16*(5119), 1–29. https://doi.org/10.3390/ijerph16245119

Skyhook Foundation. (2025). *The Skyhook Foundation*. https://skyhookfoundation.org/skyhook-foundation/#sf_camp_skyhook

Soliveres, S., der Plas, F., Manning, P., Prati, D., Gossner, M.M., Renner, S.C., Alt, F., Arndt, H., Baumgartner, V., Binkenstein, J., Birkhofer, K., Blaser, S.,… & Fischer, M. (2016). Biodiversity at multiple trophic levels is needed for ecosystem multifunctionality. *Nature*, *536*, 456–459. https://doi.org/10.1038/nature19092

State of California. (2024, May 7). *World's largest wildlife crossing on track to open by early 2026*. Governor Gavin Newsom. www.gov.ca.gov/2024/05/07/worlds-largest-wildlife-crossing…01%20is%20one,in%20Southern%20California%2C%20including%20people.

Thompson, E., Herian, M., & Rosenbaum, D. (2021, June 28). *The economic footprint and quality-of-life benefits of urban forestry in the United States*. Bureau of Business Research, University of Nebraska—Lincoln. www.arborday.org/tree-resources/economics-urban-forestry

Trust for Public Land. (n.d.). *The heat is on: A Trust for Public Land special report.* www.tpl.org/wp-content/uploads/2022/09/The-Heat-is-on_A-Trust-for-Public-Land_special-report.pdf

United Nations Economic Commission for Europe. (n.d.a). *Trees in cities challenge.* https://treesincities.unece.org/#sthash.HwOqNFph.dpbs

United Nations Economic Commission for Europe. (n.d.b). *Over 80 cities have pledged to date.* https://treesincities.unece.org/tree-o-meter#sthash.7DzYHNRT.oNXb4ThW.dpbs

United Nations Environment Programme. (2019). *Environmental rule of law: First global report.* www.unep.org/resources/assessment/environmental-rule-law-first-global-report

United Nations General Assembly. (2022, July 28). *With 161 Votes in Favour, 8 Abstentions, General Assembly Adopts Landmark Resolution Recognizing Clean, Healthy, Sustainable Environment as Human Right.* [Press release]. https://press.un.org/en/2022/ga12437.doc.htm#:~:text=With%20161…ons%2C%20the%20General,ensure%20that%20principle%20is%20upheld

U.S. Forest Service. (n.d.). *Bat pollination.* U.S. Department of Agriculture. www.fs.usda.gov/managing-land/wildflowers/pollinators/who-are-the-pollinators/bats

U.S. Forest Service. (n.d.). *North American Beaver.* U.S. Department of Agriculture. www.fs.usda.gov/detail/npnht/learningcenter/nature-science/?cid=fseprd878831

U.S. National Park Service. (2021, October 6). *Acadia's North American beaver: The ultimate keystone species.* www.nps.gov/articles/north-american-beaver-acadia.htm

Washington State Department of Fish & Wildlife. (2024a). *Wildlife habitat yards.* https://wdfw.wa.gov/species-habitats/living/habitat-at-home/wildlife-yards

Washington Department of Fish & Wildlife. (2024b). *Community science at home.* https://wdfw.wa.gov/species-habitats/living/habitat-at-home/community-science

Van Den Eeden, S.K., Browning, M.H.E.M., Becker, D.A., Shan, J., Alexeeff, S.E., Ray, G.T., Quesenberry, C.P., & Kuo, M. (2022). Association between residential green cover and direct healthcare costs in Northern California: An individual level analysis of 5 million persons. *Environment International, 163*, 1–8. https://doi.org/10.1016/j.envint.2022.107174

Empowered to Improve Planetary Health 12

> "That's how the world changes. People stand up for what they believe in".
>
> Baron Victor Adebowale[1]

OVERVIEW

The late Kenyan parliamentarian and Nobel Laureate Dr. Wangari Maathai said, "It is the people who must save the environment. It is the people that must make their leaders change. And we cannot be intimidated..." (Merton & Dater, n.d.). Dr. Maathai founded the grassroots Greenbelt Movement in 1977 where Kenyan women planted trees to combat the consequences of deforestation and diminished fresh water sources (Dater & Merton, 2008; Maathai, 2004). Through the simple act of planting trees, the Greenbelt Movement became empowered to address the larger issues of governance, environmental assault, and the disempowerment of citizens, where trees symbolized democracy (Dater & Merton, 2008; Maathai, 2004). At the time Dr. Maathai was awarded the Nobel Prize for Peace, the Greenbelt Movement planted 30 million trees in 30 years (Maathai, 2004). Dr. Maathai stressed that *we can work together* to embrace the entirety life on Earth (Dater & Merton, 2008; Maathai, 2004). This last chapter starts by reinforcing the importance of whole plant-foods nutrition as the foundational dietary pattern that supports planetary health where individuals, households, and communities can create critically important outcomes. Thereafter, recycling pitfalls will be discerned while inspiring vignettes will reinforce the power of determination in improving planetary health.

DOI: 10.1201/9781003153528-14

INDIVIDUAL AND COLLECTIVE ACTIONS FOR PLANETARY HEALTH

Households in high-income nations account for a substantial share of annual global greenhouse gas emissions (Dubois et al., 2019). Dubois et al. (2019) studied four cities in Germany, France, Norway and Sweden that included household footprint assessments and found that nearly 64% of the household footprint came from dietary intake (i.e. eating meat) and transportation (Dubois et al., 2019). Wynes and Nicholas (2017) found that at the individual level, behaviors that have the greatest impact to reduce personal greenhouse gas emissions included eating a plant-based diet, living car-free, having one less child, and avoiding air travel.

Plant-Rich Diets and Avoiding Food Waste

Noting that, "Some of the most powerful climate solutions receive comparably little attention, reminding us to widen our lens", the concept of "drawdown" denotes "the future point in time when levels of greenhouse gases in the atmosphere stop climbing and start to steadily decline" (Project Drawdown, 2020). Project Drawdown is a non-profit organization and network of scientists and researchers that have examined over 90 top sources of global emissions, practices, and technologies, analyzing and summarizing their importance as solutions to mitigate climate impacts (Frischmann & Chissell, 2021; Project Drawdown., 2020). Their analyses notes that individual and household actions can reduce total emissions by 25-30% (Frischmann & Chissell, 2021). More specifically, *plant-rich diets along with reducing food waste offer the largest opportunity* for reducing cumulative carbon emissions over the next 30 years, accounting for *the greatest impact across sectors* as seen in Figure 12.1 (Frischmann & Chissell, 2021).

Although greenhouse gas emissions (GHGE) associated with the American diet have decreased—they are still well above the necessary targets (Bassi et al., 2022). Similarly, data analyses from the United Kingdom revealed significant decreases in red and processed meat consumption although efforts must be amplified to meet climate targets (Stewart et al., 2021). Bozeman et al. (2020) utilized lifecycle analyses to compare Americans' dietary patterns to the planetary health diet discussed in Chapter 9. They too found that Americans fall short of meeting dietary guidelines for planetary health while their subanalyses revealed inequities and barriers for persons of color (Bozemen et al., 2020).

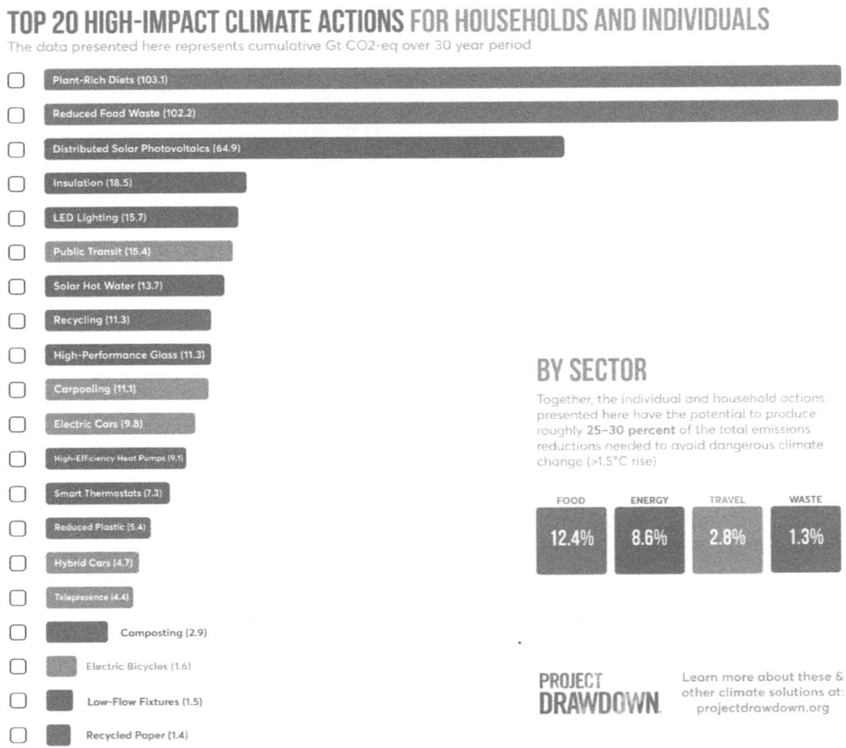

FIGURE 12.1 Project Drawdown's Top 20 Climate Actions for Households and Individuals.

Creating Climate-Friendly Meals in Schools

An impressive example of individual and collective actions that increase the consumption of plant-based meals comes from the Friends of the Earth (FOE) Climate Friendly School Food Program. This program focuses on K-12 schoolchildren where 7 billion meals are served annually in the U.S. (Friends of the Earth [FOE], n.d.a). In California alone, over a billion meals are served annually where processed foods containing animal products dominate menus (Hamerschlag et al., 2020; Stewart & Hamerschlag, 2023). Due to the impact of this program in California, there's been an impressive increase in plant-based entrees served however, they comprised only 11% of the entrees served in 2023—indicating there's more work to do (Stewart, 2024).

As California is a racially and ethnically diverse state, reflecting this diversity in school food choices has increased plant-foods consumption where staff creativity and collaboration with students has led to celebrated entrees such as Chana Masala, vegan tamales, falafel wraps, tofu Katsu bowls, Singapore curry noodles, and Cajun beans (Stewart & Hamerschlag, 2023). In the San Luis Obispo school district, an experimental menu item—the Thai Basil Lentil Burger as noted in Figure 12.2—became a favorite entrée where some ingredients were locally sourced (Friends of the Earth, FOE, n.d.b).

San Luis Coastal Unified Coastal School District Food Service Director Erin Primer remarked, "…there has been a positive response to plant-forward foods, even among meat eaters. Presenting plant-forward food as healthy, fresh and climate-friendly is an inclusive marketing strategy that everyone can be a part of" (FOE, n.d.b). Forty-one new and culturally diverse plant-based dishes were served in 2023, adding to the 102 plant-based entrees served in California's 25 largest school districts (Stewart, 2024). Importantly, menus with processed meats decreased by 25% from 2022–2023 (Stewart, 2024). Fifty-six percent (56%) of high schools and middle schools offered plant-based options daily in 2023, an increase from 36% in 2019 (Stewart, 2024).

The FOE Climate Friendly School Food Program's many successes in California have come from a wide range of partnerships and efforts such as securing full-time positions for school foodservice employees, enabling them to cook from scratch versus relying on processed and packaged foods; acquiring grant monies and updating school foodservice equipment; providing education and training; sourcing and serving local organic foods; and reducing food packaging and waste (Hamerschlag et al., 2020). Additionally, improving school food choices was cost effective compared to the standard menu items served (Hamerschlag et al., 2020). However, to rapidly advance planetary health, this program recommends sweeping policy changes nationally that includes but is not limited to reforming the United States Department of Agriculture's food programs to support unprocessed or minimally processed whole-plant foods

FIGURE 12.2 Thai Basil Lentil Burger, Friends of the Earth Climate Friendly School Program, San Luis Coastal Unified School District. Photo ©Lauren Thomas.

in school foodservice; funding nutrition education; investing in school kitchen facilities and employees; and aligning national health guidance with the foods served in schools (Stewart, 2024).

The Eastside HEAL (Healthy Eating Active Living) Zone, Riverside, California

Riverside is a large multicultural city of nearly 315,000 persons in southern California located an hour east of Los Angeles; just over an hour from the Pacific Ocean; over an hour from the native palm oases of the Coachella Valley; and less than an hour to the alpine peaks of the San Bernardino National Forest (U.S. Census Bureau, 2023). Over half of its population is identified by the U.S. Census Bureau (2023) as Hispanic or Latino and nearly 47% of the population speaks a language other than English at home. To improve health outcomes in the underserved community, a unique partnership with a diverse array of stakeholders ensued to empower residents to create outcomes that prioritize unity, health, well-being, and hope (Riverside Community Health Foundation, n.d.a). What started as an effort to combat obesity by reducing calorie consumption, increasing the consumption of healthy foods and beverages, and increasing physical activity, Riverside's Eastside HEAL (Healthy Eating Active Living) Zone has become an integral force where community residents shape positive outcomes for planetary health (City of Riverside, 2016). Their emblem is noted in Figure 12.3.

Kaiser Permanente, a founding partner of the HEAL Cities campaign invests in reshaping the physical environment to engage healthy lifestyle behaviors (Kaiser Permenente, n.d.). Additionally, this campaign incorporates public policy, individual behavior, and resident empowerment (Kaiser Permenente, n.d.). Residents are encouraged to participate in the Eastside Resident Leadership Academy training that includes but is not limited to understanding the Social Determinants of Health (SDH); safe and walkable communities; healthy food systems; leadership skills; and advancing public policy (Riverside Community Health Foundation, 2017).

Some of the Eastside HEAL Zone's important accomplishments include 1) the Healthy Living Project Youth Club where youth ages 12-19 create positive outcomes in the community 2) the Residents of Eastside Active in Leadership (REAL) program where adults create positive community outcomes 3) health ministry trainings and faith leader meetings that include health programming at churches 4) a health-focused annual Holiday Posada 5) installing a shade structure at a park 6) opening a community fitness center that is free for residents 7) restoring alleyways and creating walking corridors 8) creating art displays in the community and hanging Eastside HEAL Zone

FIGURE 12.3 Riverside, California HEAL (Healthy Eating Active Living) Zone Emblem.

banners 9) increasing the amount of locally grown produce for sale at the Al Centro Market along with offering reduced price produce bags and nutrition education for recipients 10) a children's food justice program 11) "Teen Cuisine" cooking classes focused on preparing healthy meals and snacks and 12) a Senior Fitness Fair and Expo (Riverside Community Health Foundation, n.d.b; Riverside Community Health Foundation, n.d.c).

The success of the Eastside HEAL Zone along with the additional unmet needs of the community allowed East Riverside to become a California Transformative Climate Community through a competitive process where disadvantaged communities are offered the opportunity to create important outcomes (California Strategic Growth Council, 2023). East Riverside is one of 37 grantees of this place-based, resident-driven initiative designed to reduce greenhouse gas emissions while offering health and economic benefits too (California Strategic Growth Council, 2023). Projects include: installing solar power for low-income homeowners; planting ~2,000 residential, shade, and fruit trees; replacing lawns with waterwise landscaping; increased community engagement through Community Outreach Workers and Promotoras (Spanish-speaking health leaders); workforce development training in solar panel installation and construction apprenticeships; creating affordable housing; and increasing bus infrastructure and interconnectivity while offering incentives for increased ridership (City of Riverside, n.d.). Ultimately, the far-reaching actions of the East Riverside community serves as an uplifting model where residents proudly invest in improving equity and planetary health.

RECYCLING PITFALLS

As demonstrated throughout this book, the amount of waste generated globally and the massive consumption of resources is unsustainable. Businesses and individuals in the U.S. generate the most municipal solid waste, plastic waste, food waste, and hazardous waste globally (Nichols & Smith, 2019). Seemingly, recycling is a pro-environmental behavior that is symbolized by the familiar green triangular chasing arrows "reduce-reuse-recycle". However, recycling rates in the United States are generally low where plastic recycling is altogether highly problematic, extremely limited, and ensnared by both misinformation and disinformation (Young, 2020; U.S. Environmental Protection Agency [EPA], 2020). With more convenience foods utilized, food packaging is ubiquitous and lucrative being valued at more than $10 billion in the U.S. (Shahbandeh, 2024). Both globally and nationally, the food packaging industry is expected to grow (Shahbandeh, 2024). As previously discussed, food packaging is destructive to the natural environment and threatens human health. Hence, reducing food packaging is extremely important to improve planetary health.

The U.S. Environmental Protection Agency (EPA) (2020) reports that overall, just over 32% of the solid waste generated in 2018 was recycled (U.S. Environmental Protection Agency [EPA], 2020). Only 8.7% of plastics generated that year was recycled along with 17% of aluminum and 25% of glass, respectively (EPA, 2020). Generally, glass can be recycled continuously however, glass manufacturing and recycling is not without negative externalities while the increasingly limited availability of a key raw material, sand, is concerning (Beiser, 2019; Colangelo, 2024). In fact, McDonough and Braungart (2002) made the case decades ago that products can be designed to be reused continuously *from the outset* in a "cradle-to-cradle" approach (versus a "cradle-to-grave" approach). Judicious use of the remaining resources we have left is imperative for planetary health.

Refillable Glass Beverage Containers

No threshold has been established that discerns a safe intake of alcohol consumption (Anderson et al., 2023). This vignette is not intended to promote alcohol consumption or to discredit the merits of recycling—but serves to highlight the ingenuity and determination involved in moving away from the business as usual practice of using single-use beverage containers—to efforts that reclaim the integrity of reusing a glass container many times (Gordon, 2021).

Matt Swihart is a brewmaster and owner of the Double Mountain Brewery & Cidery in Hood River, Oregon (BottleDrop Oregon Redemption Center, 2024). Concerned with serving the highest quality beverages, Swihart not only chose glass containers—but started bottling his products in 2012 in reusable glass containers that were refillable up to 25-30 times (M. Swihart, personal communication, November 19, 2024). When living in Wisconsin, Swihart used refillable glass bottles as a consumer and therefore knew it was possible however, there were hurdles and mixed-successes initially at his Oregon business such as sourcing a company to clean the bottles and pushback from retailers (M. Swihart, personal communication, November 19, 2024). Later, the Oregon Beverage Recycling Cooperative (a not-for-profit statewide operator) streamlined the process in conjunction with its bottle redemption centers and Oregon's bottle deposit legislation (BottleDrop Oregon Redemption Center, 2024). As of November 2024, Swihart calculated that his business utilized over 2.5 million refillable glass bottles (instead of 2.5 million single-use glass bottles) for a reduction of 448,864 pounds of carbon dioxide emissions (M. Swihart, personal communication, November 19, 2024).

Reusing glass is associated with substantial decreases in GHGE compared to emissions from single-use glass containers and offers lower carbon emissions compared to other containers, too (Gordon, 2021; Oregon Department of Environmental Quality, 2024). Furthermore, Oregon's Senate Bill 582 "Plastic Pollution and Recycling Modernization Act" will overhaul the state's recycling system that includes but is not limited to extended producer responsibility for packaging; improving recycling services and facilities; improving equity; and substantially decreasing environmental impact (Oregon Department of Environmental Quality, 2021).

In closing, this last chapter has demonstrated the power of individuals and their collaborative efforts to create outcomes that improve planetary health. As Williamson et al. (2018) has noted, "Solving the global climate change crisis is going to rely on, in one way or another, changing human behavior". Indeed, it will take people working together in all aspects of planetary health to resolve the multifaceted planetary crises we face. Although seemingly daunting, there's a great deal we can do every day—from establishing whole plant-foods nutrition as the preferred foundational dietary pattern—to eliminating food waste—to creating equity—and bolstering biodiversity. Despite the hurdles, we can create the important outcomes we need as a global community with unwavering determination and actions. As the spoken word artist Prince Ea conveys in his piece, *Dear Future Generations: Sorry*, "…an error does not become a mistake until you refuse to correct it" (Ea, 2015).

To end this book with the sentiments of the exemplar discussed at the beginning of this chapter, Dr. Wangari Maathai said, "In the course of history, there comes a time when humanity is called to shift to a new level of consciousness, to reach a higher moral ground. A time when we have to shed our fear and give hope to each other. That time is now" (Maathai, 2004).

NOTE

1 Gething, A. (Director). *Diana, our mother: Her life and legacy.* [Film]. HBO Documentary Films. 45:20.

REFERENCES

Anderson, B.O., Berdzuli, N., Ilbawi, A., Kestel, D., Kluge, H.P., Krech, R., Mikkelsen, B., Neufelt, M., Poznyak, V., Revke, D., Slama, S., Tello, J., & Ferreira-Borges, C. (2023). Health and cancer risks associated with low levels of alcohol consumption. *The Lancet Public Health*, *8*, e6–e8.

Bassi, C., Maysels, R., & Anex, R. (2022). Declining greenhouse gas emissions in the US diet (2003–2018): Drivers and demographic trends. *Journal of Cleaner Production*, *351*, 1–10. https://doi.org/10.1016/j.jclepro.2022.131465

Beiser, V. (2019, November 17). Why the world is running out of sand. *BBC*. www.bbc.com/future/article/20191108-why-the-world-is-running-out-of-sand

BottleDrop Oregon Redemption Center. (2024). *BottleDrop refillable bottles: Powered by OBRC.* https://bottledrop.com/craft-beverage-makers/

Bozeman, J.F., Springfield, S., & Theis, T.L. (2020). Meeting EAT-Lancet food consumption, nutritional, and environmental health standards: A U.S. case study across racial and ethnic subgroups. *Environmental Justice*, *13*(5), 160–172.

California Strategic Growth Council. (2023, December). *Transformative climate communities.* https://sgc.ca.gov/grant-programs/tcc/docs/20231218-TCC_Fact_Sheet.pdf

City of Riverside. (2016). *HEAL (Healthy Eating Active Living) Zone.* https://riversideca.gov/park_rec/programs-sports/health-wellness/heal-healthy-eating-active-living-zone

City of Riverside. (n.d.). *Transformative climate communities grant.* https://riversideca.gov/hhs/about/grants/tcc-grant-projects

Colangelo, S. (2024). Reducing the environmental footprint of glass manufacturing. *International Journal of Applied Glass Science*, *15*, 350–366. https://doi.org/10.1111/ijag.16674

Dater, A., & Merton, L. (Directors). (2008). *Taking root: The vision of Wangari Maathai.* [Film]. Marlboro Productions.

Ea, P. (2015, April 14). *Dear Future Generations: Sorry. An Apology Letter to Future Generations.* [Video]. You Tube. www.youtube.com/watch?v=eRLJscAlk1M

Dubois, G., Sovacool, B., Aall, C., Nilsson, M., Barbier, C., Herrmann, A., Bruyère, S., Andersson, C., Skold, B., Nadaud, F., Dorner, F., Moberg, K.R., Cerona, J.P., Fischer, H., Amelung, D., Baltruszewicz, M., Fischer, J., Benevise, F., Louis, V.R., & Sauerborn, R. (2019). It starts at home? Climate policies targeting household Consumption and behavioral decisions are key to low-carbon futures. *Energy Research & Social Science, 52,* 144–158. https://doi.org/10.1016/j.erss.2019.02.001

Friends of the Earth. (n.d.a). *Climate-friendly school food.* https://foe.org/projects/school-food-purchasing/

Friends of the Earth. (n.d.b). *How a rockstar lentil burger inspired more plant-forward school food.* https://foe.org/resources/slo-lentil-burger-study/

Frischman, C., & Chissell, C. (2021, October 27). The powerful role of household actions in solving climate change. *Project Drawdown.* https://drawdown.org/insights/the-powerful-role-of-household-actions-in-solving-climate-change

Gordon, M. (2021). Reuse wins: The environmental, economic, and business case for transitioning from single-use to reusable in food service. *Upstream.* https://drive.google.com/file/d/1DTKK54rLhRRQGbX40pG891f3CYevpDJg/view

Hamerschlag, K., & Ardnt, S. (2020). *Organic, plant-forward, scratch cooked meals: A California case study.* www.consciouskitchen.org/wpcontent/uploads/2021/01/CK_CaseStudy_CompleteReport-2.pdf

Kaiser Permanente. (n.d.). *Healthy Neighborhoods.* https://about.kaiserpermanente.org/commitments-and-impact/healthy-communities/improving-community-conditions/healthy-neighborhoods

Maathai, W. (2004, December 10). *Nobel lecture.* www.nobelprize.org/prizes/peace/2004/maathai/lecture/

McDonough, W., & Braungart, M. (2002). *Cradle to cradle: Remaking the way we make things.* North Point Press.

Merton, L., & Dater, A. (n.d.). *Planting ideas: International action guide.* http://takingrootfilm.com/wp-content/uploads/2015/08/Taking-RootInternationalActionGuide.pdf

Nichols, W., & Smith, N. (2019). *Waste generation and recycling indices 2019 overview and findings.* Verisk Maplecroft. www.circularonline.co.uk/wp-content/uploads/2019/07/Verisk_Maplecroft_Waste_Generation_Index_Overview_2019.pdf

Oregon Department of Environmental Quality. (2024, May). *2022 Oregon material recovery and waste generation rates report.* www.oregon.gov/deq/recycling/Documents/2022MRWGRatesReport.pdf

Oregon Department of Environmental Quality. (2021, August 10). *Oregon adopts groundbreaking Plastic Pollution and Recycling Modernization Act.* [Press release]. https://apps.oregon.gov/oregon-newsroom/OR/DEQ/Posts/Post/oregon-ad...undbreaking-plastic-pollution-and-recycling-modernization-act-42917

Project Drawdown. (2020). *The drawdown review: Climate solutions for a new decade.* https://drawdown.org/sites/default/files/pdfs/TheDrawdownReview–2020–Download.pdf

Riverside Community Health Foundation. (n.d.a). *Eastside HEAL zone.* https://rchf.org/programs/eastside-heal-zone/

Riverside Community Health Foundation. (n.d.b). *Eastside HEAL zone phase II impact report.*

Riverside Community Health Foundation. (n.d.c). *Eastside HEAL zone phase I impact report.*

Riverside Community Health Foundation. (2017). *Meet and greet.* https://rchf.org/wp-content/uploads/2017/05/Meet-and-Greet-2017.pdf

Shahbandeh, M. (2024). *Food packaging industry worldwide- statistics & facts.* Statista. www.statista.com/topics/10806/food-packaging-industry/#topicOverview

Stewart, C. (2021). Trends in UK meat consumption: Analysis of data from years 1–11 (2008–09 to 2018–19) of the National Diet and Nutrition Survey rolling programme. *The Lancet Planetary Health*, *5*, e699–e708.

Stewart, N. (2024). *Plant-based trends in California's school lunches: A 5-Year assessment.* https://foe.org/wp-content/uploads/2024/09/Plant-Based-Trends-Report_v5.pdf

Stewart, N., & Hamerschlag, K. (2023). *Plant-based trends in California's school lunches: A progress report on climate-friendly school food.* https://foe.org/wp-content/uploads/2023/09/Plant-Based-Trends-in-California-School-Lunches_Final.pdf

U.S. Census Bureau. (2023). *Riverside city, California.* https://data.census.gov/profile/Riverside_city,_California?g=160XX00US0662000

U.S. Environmental Protection Agency. (2020). *Advancing sustainable materials management: 2018 Fact Sheet.* www.epa.gov/sites/default/files/2020-11/documents/2018_ff_fact_sheet.pdf

Williamson, K., Satre-Meloy, A., Velasco, K., & Green, K. (2018). *Climate change needs behavior change: Making the case for behavioral solutions to reduce global warming.* https://rare.org/wp-content/uploads/2019/02/2018-CCNBC-Report.pdf

Wynes, S., & Nicholas, K.A. (2017). The climate mitigation gap: Education and government recommendations miss the most effective individual actions. *Environmental Research Letters*, *12*, 1–9. https://doi.org/10.1088/1748-9326/aa7541

Young, R. (Writer), & Huffman, B.E., et al. (Directors). (2020, March 31). Plastic Wars (Season 2020, Episode 14) [Tv series episode]. In R. Young et al. (Executive Producers), *Frontline*. Public Broadcasting Corporation.

Appendix

USING ESTABLISHED OBSERVANCES TO ADVANCE PLANETARY HEALTH

Celebrating established observances reinforces our commitment to improving planetary health outcomes while encouraging action through goodwill and creativity. Using World Pulses Day as an example, host a cook-off—offer a taste-test—or share recipes and fact sheets that highlight the many co-benefits that pulses offer which are noted in **Figure A.1**.

Lastly, **Table A.1** offers suggestions for observances that advance nutrition, nature, equity, and planetary health.

FIGURE A.1 World Pulses Day, February 10. Food and Agriculture Organization.

TABLE A.1 Promoting Sustained Actions for Planetary Health

DATE	*ESTABLISHED OBSERVANCE*
February 2	World Wetlands Day
February 4	International Day for Human Fraternity
February 10	World Pulses Day
February 20	World Day for Social Justice
March 1	World Seagrass Day
March 3	World Wildlife Day | African Environment Day | Wangari Maathai Day
March 8	International Women's Day
March 20	International Day of Happiness
March 21	International Day of Forests
March 21	International Day for Eliminating Racial Discrimination
March 22	World Water Day
March 23	World Meteorological Day
March 25	International Day of the Victims of Slavery and the Transatlantic Slave Trade
March 30	International Day of Zero Waste
April 1-30	Native Plant Month
April 7	World Health Day
April 7	International Beaver Day
April, 3rd Week	California Native Plant Week
April 22	International Mother Earth Day
May 1-31	No Meat May
May 5	California Native Plant Day
May 12	International Day of Plant Health
May 14	World Migratory Bird Day
3rd Friday	Endangered Species Day
May 18-24	Lifestyle Medicine Week
May 20	World Bee Day
May 22	International Day for Biological Diversity (Biodiversity)
June 1	National Trails Day
June 1-9	California Invasive Species Action Week
June 3	World Bicycle Day
June 8	World Ocean Day
June 15 or 16	Be Love Day, The King Center, Atlanta, Georgia, USA
June 16	World Refill Day
June 17	World Day to Combat Desertification and Drought
June 18	Sustainable Gastronomy Day
July 6	World Zoonoses Day
July 30	International Day of Friendship

TABLE A.1 (Continued)

DATE	*ESTABLISHED OBSERVANCE*
August 9	International Day of the World's Indigenous Peoples
September 7	International Day for Clean Air and Blue Skies
September 16	International Day for the Preservation of the Ozone Layer
September 21	International Day of Peace
September 29	International Day for Awareness of Food Loss and Waste
October 2	International Day for Non-Violence
October 4	World Habitat Day
October 16	World Food Day
November 3	One Health Day
November 6	International Day for Preventing the Exploitation of the Environment in War and Armed Conflict
November 13	World Antimicrobial Awareness Week
November 14	World Diabetes Day
December 5	World Soil Day
December 10	Human Rights Day
December 11	International Mountain Day
December 11	Right to a Sustainable Environment

Index

Note: Page numbers in *italic* refers to Figures.

FSC
www.fsc.org
MIX
Papier | Fördert gute Waldnutzung
FSC® C083411

Kolibri 360 GmbH
Ferdinand-Jühlke-Straße 7
99095 Erfurt, Deutschland
produktsicherheit@kolibri360.de